Student Study Guide to
Accompany Wendy J. Steinberg's

Statistics
Alive! 2 EDITION

Student Study Guide to Accompany Wendy J. Steinberg's

Statistics

Alive! 2 EDITION

prepared by
Matthew Price
Georgia State University

Los Angeles | London | New Delhi
Singapore | Washington DC

For information:

SAGE Publications, Inc.
2455 Teller Road
Thousand Oaks, California 91320
E-mail: order@sagepub.com

SAGE Publications Ltd.
1 Oliver's Yard
55 City Road
London EC1Y 1SP
United Kingdom

SAGE Publications India Pvt. Ltd.
B 1/I 1 Mohan Cooperative
    Industrial Area
Mathura Road, New Delhi 110 044
India

SAGE Publications Asia-Pacific
    Pte. Ltd.
33 Pekin Street #02-01
Far East Square
Singapore 048763

Printed in the United States of America

ISBN 9781412994286

This book is printed on acid-free paper.

10   11   12   13   14   10 9 8 7 6 5 4 3 2 1

Acquisitions Editor:	Vicki Knight
Associate Editor:	Lauren Habib
Proofreader:	A. J. Sobczak
Cover Designer:	Bryan Fishman

Module 1

Learning Objectives

- Become familiar with common statistical terms and symbols

- Review common arithmetic rules and functions

Module Summary

- Statistics is not a difficult subject, but it can seem that way. One of the difficult aspects of statistics is that it uses many different terms, which can be confusing when you are first learning the material. You should become familiar with all of the terms as you read the textbook to help you better understand the material. Another aspect that is commonly thought of as difficult is the mathematical component of statistics. However, the actual math that will be used for your introduction to statistics is relatively simple. In fact, in this text book the hardest mathematical operations are exponents (X^2) and square roots ($\sqrt{}$). The difficult part of learning this material will be understanding the logic behind the math. To better grasp this logic, it is important for you to become fluent with the different terms.

- Here is a brief list of common statistical terms and their definitions.

    - Case/Subject/Participant: a unit of study. The term *participant* is preferred by the American Psychological Association (APA).

    - Sample: A group of subjects in a study that are part of a larger group.

    - Population: A large group from which samples are drawn. We use samples to learn about populations.

    - Statistic: A number that summarizes a sample.

    - Parameter: A number that summarizes a population.

1

- Variable: A measurement that can vary from person to person in a sample. You can consider the height and weight of 5 different people to be variables because they will be different (vary) from person to person.

- Constant: A measurement that remains the same for all cases.

- Uppercase letters: These are used to represent variables.

- Bar over a letter: This represents an average.

- M: The mean of a sample.

- p: The probability of an event occurring.

- q: The probability that an event will not occur.

- N, n: The *number* of cases. N refers to the number of cases in a population, whereas n refers to the number of cases in a sample.

- Subscripts: Refer to a specific case. X_1 is the first case's score on variable X.

- Wavy parallel lines: The symbol $\approx$ indicates an approximation, as opposed to $=$, which indicates a definite amount.

- $<$ and $>$: Less than a value and more than a value.

- Summation: Adding up all scores for a particular variable.

- Reciprocal: Dividing one by the number. The reciprocal of 7 is 1/7.

- Exponent (superscripted number): Tells you to multiple a number by itself as many times as the exponent, or superscripted number.

- Radical sign: Take the square root of the number under the radical sign.

- Here are a few simples rules to keep in mind as you work through computations:

    - Dividing a number is equivalent to multiplying that number by its reciprocal.

- When multiplying negative and positive numbers, the product will always be negative. When multiplying two negative numbers, the product will always be positive.

- Fractions, decimals, and percentages are different ways to represent the same amount: $1/4 = .25 = 25\%$.

- You can expect many of your mathematical operations to have multiple decimal places. It is common in the social sciences to round all work to three decimal places and the final answer to two decimal places.

- When rounding, values of 5 or greater are rounded up, meaning that 1 is added to the previous digit; 0.56 becomes 0.6. Alternatively, values of 4 or lower are rounded down, meaning the previous digit remains the same; 0.54 becomes 0.5.

- Always remember the order of operations in math. These are (1) work in parentheses, (2) exponents and square roots, (3) multiplication, (4) division, and (5) addition and subtraction. The acronym PEMDAS can help you to remember this order.

- Finally, remember that you may need to re-order an equation to solve for an unknown value. Remember the basic rules of algebra, which state that what is done to one side of an equation (equals sign) must be down to the other. For example, if $12 = 4X + 4$, subtracting 4 from both sides to give you $12 - 4 = 4X$. Both sides can then be divided by 4, which provides the answer of $X = 2$.

- $\Sigma(X)$ is the symbol for summation. It indicates that you should sum all of the values of X.

Computational Exercises

1. $(3)(6) + 2 =$

2. $\dfrac{(14+13)}{3} =$

3. $\dfrac{(3+7)2}{4} =$

4. Round your answer to two decimal places: $\dfrac{(9+13)6}{8} =$

5. Round the following numbers to two decimal places:

   a. 4.57813; b. 3.2143; c. 8.9431

6. What is the reciprocal of the following:

   a. 4; b. 3; c. 2; d. 2/3

7. Complete the following chart:

Fraction	Percent	Decimal
9/45		
	92%	
		0.35

8. Solve the following equations, remembering the rules of the order of operations:

   a. $(12+45) - (54-9) =$

   b. $\dfrac{12+56}{4} + \dfrac{12}{2+4} =$

   c. $\dfrac{12+56+12}{4+2+4} =$

   d. $(4+3)^2 - (20-7) =$

   e. $\sqrt{5+11} + \left(\dfrac{5+4}{3}\right)^2 =$

9. Solve for X in the following equations:

   a. $X + 4 = 9$

   b. $4X - 7 = 2$

   c. $2X - (3+4)^2 = 83$

10. Solve for X in the following equations:

a. $aX + b = c$

c. $\dfrac{b(X - a)}{(c + d)} = Z$

b. $\dfrac{aX - b}{c} = Z$

11. Expand and solve the following expressions if $X_1 = 2$; $X_2 = 4$; $X_3 = 9$:

a. $\sum(X) =$

d. $\sum(2X) =$

b. $\left(\sum(X)\right)^2 =$

e. $2\sum(X) =$

c. $\sum(X^2) =$

12. Expand the following binomials:

a. $2(a + b) =$

c. $3(a + b)^2 =$

b. $(a + b)^2 =$

Computational Answers

1. 20

2. 9

3. 5

4. 16.5

5. a. 4.58; b. 3.21; c. 8.94

6. a. 1/4; b. 1/3; c. 1/2; d. 3/2

7. Complete the following chart:

Fraction	Percent	Decimal
9/45	20%	0.2
23/25	92%	0.92
7/20	35%	0.35

5

8. Solve the following equations, remembering the rules of the order of operations:

   a. 12; b. 19; c. 8; d. 36; e. 13

9. Solve for X in the following equations:

   a. $X = 5$; b. $X = 2.25$; c. $X = 66$

10. Solve for X in the following equations:

    a. $X = \dfrac{c-b}{a}$; b. $X = \dfrac{Zc+b}{a}$; c. $X = \dfrac{Z(c+d)+ba}{b}$

11. Expand and solve the following expressions if $X_1 = 2$; $X_2 = 4$; $X_3 = 9$:

    a. 15; b. 225; c. 101; d. 30; e. 30

12. Expand the following binomials:

    a. $2a + 2b$; b. $a^2 + 2ab + b^2$; c. $3a^2 + 6ab + 3b^2$

True/False Questions

1. Statistics are used to summarize populations, whereas parameters are used to summarize samples.

2. Both X-bar and M are symbols for means.

3. p and q are related in that as one increases the other decreases.

4. The absolute value of -8 is -8.

5. In doing a complicated math problem, you should complete work in parentheses first.

6. 6.42338139 rounded to two decimal places is 6.42.

7. $\Sigma(X + Y)$ is the same as $\Sigma X + \Sigma Y$.

True/False Answers

1. False   2. True   3. True

4. False   6. True

5. True   7. True

Short-Answer Questions

1. Describe how a sample and a population differ and how they are related.

2. What is precision in mathematical calculation? How precise should you be when working on a question in statistics? When stating your answer?

3. What is the correct order in which mathematical operations must be conducted?

4. What, if any, is the difference between a case and a subject or participant?

Answers

1. Samples are small subgroups of populations. Populations consist of larger groups that contain all of the individuals in your area of interest. They are related in that samples are derived from populations and are used to infer about populations.

2. Precision is the exactness of the calculation, or how many decimal places are used. In social science statistics, we generally use three decimal places when working through a problem and present the final answer rounded to two decimal places.

3. Calculations within parentheses, exponents, multiplication, division, addition, and subtraction.

4. A case represents an individual unit of study. A subject represents an individual human being in a study.

Multiple-Choice Questions

1. If $p = .45$, then $q =$

   a. .45

   b. .65

   c. .55

   d. 1

2. The average child in your area has a height of 22 inches. What is the appropriate symbol to represent this measurement?

   a. N

   b. n

   c. M

   d. X

3. What is another way to express 4/5?

   a. .7

   b. 45%

   c. .9

   d. 80%

4. What is another way to express $\approx .67$?

   a. 4/7

   b. 4/6

   c. 65%

   d. 46%

5. According to the conventions of the social sciences, what is the solution: $4.541 + 6.327$?

   a. 10.868

   b. 10.87

   c. 10.9

   d. 10

6. What is another method to express $\Sigma(3X + 4Y)$

   a. $(12)\Sigma(X + Y)^2$

   b. $\Sigma(3X) + \Sigma(4Y)$

   c. $3\Sigma(X) + 4\Sigma(Y)$

   d. $\Sigma(3X)^2 + \Sigma(4Y)^2$

7. Solve for Z: $15 = \dfrac{Z(12 + 8)}{5}$

   a. $Z = 20$

   b. $Z = 3.75$

   c. $Z = 4.25$

   d. $Z = 10.50$

8. Solve for X: $20X - 18 = 6(X + 4)$

    a. $X = 4$ c. $X = 12$

    b. $X = 5.5$ d. $X = 3$

Multiple-Choice Answers

1. C   5. B

2. C   6. B

3. D   7. B

4. B   8. D

Module Quiz

These questions are designed to touch upon the most crucial concepts of the module.

1. Expand and solve the following expressions if $X_1 = 10$; $X_2 = 5$; $Y_1 = 3$; $Y_2 = 2$

    a. $\sum (X + Y) =$ c. $\sum (X + Y)^2 =$ e. $3\sum (X + 3Y) =$

    b. $\sum (X^2 + Y^2) =$ d. $\sum (2X + 3Y) =$

2. Complete the following chart:

Fraction	Percent	Decimal
$\approx 6/9$		
	43%	
		0.7391

3. Solve for X in the following equations:

    a. $X + 15 = 25$ d. $\dfrac{12X + 18}{10} = 132$

    b. $2X + 3 = 9$

    c. $6X - 7 = 41$ e. $\dfrac{12 - 3}{2} = 45X$

9

4. You are interested in studying the mating habits of the remaining 4,000 chimpanzees in the wild. However, you are unable to find all 4,000 chimps and so you observe 200. In this example, how many chimps are in the population, in the sample, and in a case?

5. According to the conventions of the social sciences, what is the solution: 9.87532 + 10.78672?

Quiz Answers

1.

a. 20	c. 218	e. 90
b. 138	d. 45	

2. <u>Fraction</u>   <u>Percent</u>   <u>Decimal</u>

$\approx 6/9$	67%	0.67
43/100	43%	0.43
7391/10,000	73.91%	0.7391

3. Solve for X in the following equations:

a. $X = 10$	c. $X = 8$	e. $X = 0.10$
b. $X = 3$	d. $X = 108.50$	

4. The 4,000 chimpanzees are a population and the 200 are a sample, and 1 chimp would be a case.

5. 20.66

Module 2

Learning Objectives

- Classify data according to their level of measurement

- Distinguish between discrete and continuous scores

- Establish real limits for continuously scored data

Module Summary

- Statistics are used to investigate variables. Just as there are different types of variables (age, height, gender, occupation, etc.), there are different scales in which variables can be measured. Measurement refers to the value that is assigned to a specific trait. The meaning that you assign to a certain measurement depends on the scale of measurement that was used. In statistics, there are four types of scales of measurement: nominal, ordinal, interval, and ratio.

- A *nominal scale* is one that is divided into distinct categories. Additionally, the categories are not ranked; one category is not higher or lower than the next. Favorite color would be an example of a nominal scale. My favorite color is blue, whereas yours may be red. Personal opinion aside, my preference for blue is no better or worse than your preference for red.

- An *ordinal scale* is one that ranks cases in order, but the precise difference between two cases is unknown. For example, you might classify one person as attractive and another as gorgeous. You know from these descriptors that gorgeous is above attractive, but you cannot be certain of the exact amount of improvement from one rank to the next.

- An *interval scale* is similar to an ordinal scale in that individuals are ranked in order, but in contrast to an ordinal scale, the precise difference between individuals is known. There is, however, no *true zero*. A true zero means that it is possible to have a complete absence of the variable being measured. For example, the Fahrenheit temperature scale is an interval scale. When comparing 36 degrees and 12 degrees, you can be certain that there is a 24-degree difference between the two temperatures. You cannot, however, ever have no Fahrenheit temperature. Most measures of mental ability or traits (intelligence, extraversion, etc.) also are interval scale.

- A *ratio scale* is similar to an interval scale but possesses a true zero. Again, a true zero means it is possible to have a complete absence of the variable being measured. The amount of money in your wallet is a ratio scale. This is because it is possible for you to have no money, regardless of the amount money that you actually do have.

- Variables can also be considered either continuous or discrete. *Continuous variables* have values that can fall anywhere on the scale, including between two adjacent values. An example of a continuous variable would be time (as specified by minutes and seconds). You could have a value 1 minute and 3 seconds. In comparison, *discrete variables* have values that cannot fall between adjacent values. An example of a discrete variable would be the number of students in your class, because there cannot be less than a full student, or a fraction of a student.

- Continuous variables are defined by their real limits. A *real limit* is defined as +/– .5 of the score's unit. This makes each score contiguous with the next score. Thus, for a score of 10, when the scores are scaled in units of 1, the *upper real limit* would be 10.5 and the *lower real limit* would be 9.5. For a score of 10 when scores are scaled in units of 5, the upper real limit would be 12.5 and the lower real limit would be 7.5. Note that an observed score will never fall at the real limit.

True/False Questions

1. The scale of measurement for the favorite animals of your classmates would be an ordinal scale.

2. In an interval scale, there is an equal amount of distance between adjacent ranks and there is a true zero.

3. You are attending a swim competition and notice that the swimmers are ranked by their time. Their ranking is an ordinal scale.

4. (Refer to question 3) The ranking through the use of their time is an interval scale.

5. Flavors of ice cream would be a discrete and ordinal scale.

6. The number of trumpet players in an orchestra is a discrete variable.

7. The distance that you can throw is a football is a continuous variable.

8. For a scale of 4 when scores in single units, the real limits are 3 and 5.

True/False Answers

1. False   4. False   7. True

2. False   5. False   8. False

3. True   6. True

Short-Answer Questions

1. What scale of measurement best classifies the following:

    a. Number of calories in a candy bar

    b. Colors in a bouquet of flowers

    c. Elevation in reference to sea level

2. What is the highest scale of measurement for the following:

   a. Number of available TV channels

   b. Number of pets you can own

   c. Responses to a questionnaire that ranges from –5 to 5

3. What is a nominal scale?

4. What are the limitations of a nominal scale?

5. What is the difference between an interval and a ratio scale?

6. What is meant by a ratio scale having an absolute zero?

7. What advantage does an interval scale have in comparison to an ordinal scale?

8. Distinguish between a continuous and a discrete variable.

9. Jamal proudly states that he scored twice as high as Rene on a test and therefore is twice as accomplished as her in the subject. What scale must the test use in order for Jamal's claim to be true?

10. You are conducting a study on social anxiety in males and females. You decide to classify your subjects as having either social anxiety or not having social anxiety. You then give them a question with a rating scale of –7 (very anxious) to 7 (very calm). They are then placed in a social situation, and the amount of time it takes for them to leave the situation is recorded. Identify the four variables in this study and the scales of measurement that each is using.

Answers

1. a. Ratio; b. Nominal; c. Ratio.

2. a. Ratio; b. Ratio; c. Interval or Ordinal (per debate on measurement).

3. A nominal scale is one that classifies categories.

4. Nominal scales are unable to explain rank or order.

5. An interval scale lacks a true zero point, whereas a ratio has a true zero point. The presence of a true zero indicates that it is possible to have a complete absence of the trait.

6. It means that it is possible to have a complete lack of the trait being studied.

7. An interval scale enables you to measure the exact distance between two scores.

8. Continuous variables have values that can fall anywhere on the scale. Discrete variables have distinct values, and a value cannot fall between the values.

9. A ratio scale.

10. Nominal – Male/Female; Nominal – Social anxiety / Not socially anxious; Interval or ordinal – Rating scale; Ratio – Time in a social encounter.

Multiple-Choice Questions

1. If you were interested in measuring the salary of teachers, what is the highest order scale you could use?

   a. Nominal   c. Interval

   b. Ordinal   d. Ratio

2. You are interested in studying depression. You obtain a sample of 100 people and divide them into the following categories: Depressed and Not depressed. These categories are an example of

   a. A discrete variable with an ordinal scale

   b. A continuous variable with an ordinal scale

   c. A discrete variable with a nominal scale

   d. A continuous variable with a nominal scale

3. You are interested in measuring the sizes of amoebas; however, the ruler on your microscope has broken and you are no longer able to measure them in a proper unit of measurement. As an alternative, you decide to classify them as small, medium, and large. It would be impossible for an amoeba to fall between categories. This scale is an example of

    a. A discrete variable with an ordinal scale

    b. A continuous variable with an ordinal scale

    c. A discrete variable with an interval scale

    d. A continuous variable with an interval scale

4. You obtain a score of 110 on a statistics aptitude test and scores are reported in whole points. Your upper real limit (UL) and lower real limit (LL) are:

    a. UL = 110, LL = 109.5   c. UL = 110.5, LL = 109.5

    b. UL = 110.5, LL = 109   d. UL = 111, LL = 109

5. The real limits of a scale that ranges from 1 to 7 and are reported in whole points are

    a. UL = 1, LL = 7   c. UL = 0, LL = 8

    b. UL = 1.5, LL = 6.5   d. UL = .5, LL = 7.5

6. You have created a scale measuring alcohol abuse that uses the following scale: A lot, A little, Not much, None. Your scale is:

    a. Ordinal   d. Could be Ordinal or Interval

    b. Interval   as per the debate on

    c. Ratio   measurement.

7. A chef is developing a feedback card to determine if people like his food. He creates a card that provides people with an option to check a box representing that they either enjoyed the food or disliked the food. What type of scale is this?

   a. Nominal   c. Interval

   b. Ordinal   d. Ratio

8. (Refer to question 7) The chef is upset that almost 50% of those that completed the cards checked the "disliked" box. He decides to give his customers more options to choose from to getter a better understanding of their opinion. His new scale ranges from –5 to +5. What type of scale is this?

   a. Nominal   d. Ordinal or Interval, per the

   b. Ordinal   debate on measurement

   c. Interval   e. Ratio

Multiple-Choice Answers

1. D   5. D

2. C   6. D

3. A   7. A

4. C   8. D

Module Quiz

1. In a horse race, Johnny Smith place first, while Sandy Jones places fourth. Is Johnny four times as fast as Sandy? Why or why not?

2. You are interested in obtaining the following information from the class. List the best scale of measurement for each:

   a. Favorite movie

b. Grade in last math class

c. Self rating on a 0 to 10 scale of confidence for statistics

3. Which scale would be considered the lowest level of measurement?

    a. Nominal

    b. Ordinal

    c. Interval

    d. Ratio

4. An ordinal scale

    a. Ranks scores in order on a continuous scale

    b. Places scores in discrete ranked values without equal distance between the values for the trait being measured

    c. Ranks scores in order, and the distance between adjacent ranks is equal

    d. Possesses a true zero point

5. You are asked by a professor to develop a grading scheme for his or her next class. Develop a separate grading scheme for a nominal scale, an ordinal scale, an interval scale, and a ratio scale.

Quiz Answers

1. He is not four times as fast. You cannot determine that type of information from an ordinal scale of measurement.

2. a. Nominal; b. Interval; c. Ordinal or interval, per debate on measurement.

3. A

4. B

5. Answers will vary.

Module 3

Learning Objectives

- Convert scores to frequencies: simple, cumulative, relative, cumulative relative, and grouped

- Display scores in frequency tables

- Find percentiles and percentile ranks from tabled data

Module Summary

- *Frequency tables* present data in an organized manner that enables specific information to be retrieved efficiently. Through the use of a frequency table, you could easily obtain an estimate of how many participants obtained a specific score, as well as determine how many participants scored above and below that score. All frequency tables follow a similar format. The left displays all possible values and the right displays the frequency, or how many people obtained a specific value. There are three additional columns that you could add to the frequency table to provide you with more information.

- The first additional column creates a cumulative frequency table. A *cumulative frequency table* displays how many scores fall at or below (or possibly above) a specific value. If you were to create a frequency table for test grades, this table would enable you to determine how many students were above your test grade or below your test grade.

- A *relative frequency or percentage table* tells you the proportion (percentage) of the total sample that obtained a specific score. Let's say you have a data set with 10 numbers ranging from 1 to 5, and 3 of those numbers are 4's. The relative frequency of 4's would be 30%. This is done by finding dividing the frequency of a specific score by the amount of scores in the data set and then multiplying the quotient by 100. Because relative frequencies are percentages, they must add up 100%.

- *Cumulative relative frequency or cumulative percentage tables* provide you with the percentage of scores above or below a specific value. These are created by first finding the relative frequency of each value. Then the relative frequency of the lowest score is added to the relative frequency of the next highest score. Repeat this step until you reach the highest score, which should have a cumulative relative frequency of 100%.

- When you have a large number of scores, it can be helpful to group your score into intervals when creating a frequency table (imagine listing all the possible values for the SAT, which has a scale of 0-1600!). When creating a grouped frequency table, all of the intervals should be equal in size. There are no standard rules for determining when data should be grouped or the size of each interval. The criterion is ease of interpretation.

- Cumulative relative frequencies are sometimes also referred to as *percentile ranks*. Percentile rank indicates the percentage of scores falling at or below a specific score. If you obtain a score of 85 on your next test, which has a cumulative frequency of .94 percentile rank, you can be certain that you did better than 94% of your class.

- However, since multiple scores can occur at a specified percentile rank (there may have been 6 students with a score of 85), your percentile rank provides only an estimate of your rank. To determine the precise percentile rank, you need to spread that rank across all the persons with that specific score. This is done by using the UL and LL with the following formula:

$$PR = \left(\frac{cum\ f_{LL} + (f_i / i)(X - LL)}{N} \right) 100$$

- After using the above formula, assume that the amount of scores that fall within the real limits of the score (84.5 – 85.5) are evenly distributed. To find the precise percentile rank, divide the amount of scores at that interval by the proportion (percentage) of scores in that interval and add that amount to the percentage of scores below the interval. If 90% of the scores fell below 84.5 and 4% of the scores fell between 84.5 and 85.5, you can be certain the true percentile rank of your score was 92.0.

- Alternatively, you may be interested in determining the score that corresponds to a specific percentile rank. This can be done using the following formula:

$$X_{PR} = LL + (i / f_i)(cum\ f_{UL} - cum\ f_{LL})(.5)$$

Computational Exercises

Here are the scores of 15 freshman students rating their confidence they will do well in statistics on a 1-10 scale. Use these data for questions 1-8:

5	10	10
7	4	4
3	2	10
2	1	1
2	10	9

1. Arrange the scores into a frequency table in descending order. How many students ranked their confidence as a 5? As a 6? As a 4?

2. Add a column to the table you created for question 1 to show the cumulative frequency of the scores. How many students ranked their confidence as less than 7? As greater than 4? As less than 10?

3. Add a column to the table you created for question #1 that shows a relative frequency for each score. What percentage of students ranked their confidence as a 3? As an 8?

4. Add a column to the table that you created for question #1 that shows the cumulative relative frequency.

5. What is the percentile rank for a person who rated himself or herself at a 5?

6. What is the exact percentile rank for a person who rated himself or herself at a 4? Use the formula.

7. What score falls at the 40th percentile rank?

8. What score falls at the 33rd percentile rank?

Computational Answers

1.

X	f
10	4
9	1
8	0
7	1
6	0
5	1
4	2
3	1
2	3
1	2

Number at 5 = 1; Number at 6 = 0; Number at 4 = 2.

2.

X	f	Cum f
10	4	15
9	1	11
8	0	10
7	1	10
6	0	9
5	1	9
4	2	8
3	1	6
2	3	5
1	2	2

Less than 7 = 9; as greater than 4 = 7; Less than 10 = 11.

3.

X	f	Cum f	Rel f
10	4	15	26.67%
9	1	11	6.67%
8	0	10	0.00%
7	1	10	6.67%
6	0	9	0.00%
5	1	9	6.67%
4	2	8	13.33%
3	1	6	6.67%
2	3	5	20.00%
1	2	2	13.33%

4.

X	f	Cum f	Rel f	Cum Rel f
10	4	15	26.67%	100.00%
9	1	11	6.67%	73.33%
8	0	10	0.00%	66.67%
7	1	10	6.67%	66.67%
6	0	9	0.00%	60.00%
5	1	9	6.67%	60.00%
4	2	8	13.33%	53.33%
3	1	6	6.67%	40.00%
2	3	5	20.00%	33.33%
1	2	2	13.33%	13.33%

5. 60%.

6. $PR = \left(\dfrac{6 + (2/1)(4 - 3.5)}{15} \right) 100 = 46.67\%$.

7. A score of 3.

8. $X_{PR} = 1.5 + (1/3)(5 - 2)(.5)$. A score of 2.

True/False Questions

1. In creating a frequency table, you do not need to list scores that have a frequency of zero.

2. Frequency tables are used to organize information more efficiently.

3. A cumulative frequency of 10 means that there are 10 scores below this particular score.

4. Relative frequencies are the proportion of scores at or below a specific score.

5. In a sample containing $n = 20$ participants, 6 of the participants obtained a score of 12 on a measure. The relative frequency for the score of 12 for this sample is 30%.

6. All the relative frequencies for a sample must sum to 100%.

7. The cumulative relative frequency of the lowest score in a data set is always 0%.

8. When creating a grouped frequency table, you should have a minimum of 5 intervals.

True/False Answers

1. False	4. False	7. False
2. True	5. True	8. True
3. False	6. True	

Short-Answer Questions

1. What are the advantages of organizing data in a frequency table as opposed to viewing them as raw scores?

2. You are creating a frequency table for a test anxiety scale that ranges from 1 to 10. After administering the test to 25 people, you notice that no one scored a 5 or an 8. How many intervals should your scale have?

3. If you wanted to determine how many scores were in a data set, which frequency table column would provide this information most efficiently?

4. What does a relative frequency tell you?

5. You are checking the grade of your last English test and notice that your professor provided you with a cumulative relative frequency table as well. You notice that cumulative relative frequency of your score was .59. What does this mean?

6. When should you considering making a grouped frequency table as opposed to a regular frequency table?

7. What is a percentile rank?

Answers

1. The data are organized in an efficient manner such that you can easily determine how many of any particular score are in the data set. When viewing raw scores, especially those out of order, it can be difficult to determine this information.

2. It should still have 10 intervals.

3. The cumulative frequency column.

4. The proportion of the sample that obtained a particular score.

5. 59% of the class scored the same or lower than you and 41% of the class scored higher.

6. There is no standard rule for creating a grouped frequency table. However, you should consider using this when the scale for your data set is so large that it would be cumbersome to use a standard frequency table.

7. A percentile rank is the percentage or proportion of cases falling at or below a specific score.

Multiple-Choice Questions

The following scores represent the amount fear (scored on a scale of 1-15) experienced when going through a haunted house on Halloween. Use the following table for Questions 1-6.

Score	Frequency
15	2
14	5
13	8
12	9
11	12
10	15
9	15
8	16
7	18
6	20
5	21
4	23
3	24
2	15
1	10

1. What is the cumulative frequency (starting from the bottom) at a fear of 5?

    a. 21   c. 93

    b. 72   d. 103

2. How many people visited the haunted house this past Halloween?

    a. 123   c. 15

    b. 213   d. 120

3. What is the approximate relative frequency of people that providing a rating of 10?

    a. 0.07   c. 0.05

    b. 0.70   d. 0.50

4. What is the cumulative relative frequency (starting from the bottom) of a rating of 15?

    a. 0.10   c. 0.98

    b. 0.4   d. 1.0

5. If you were to draw a relative frequency curve of this data, how would you describe the shape of this distribution?

    a. Symmetrical   c. Negatively skewed

    b. Positively skewed   d. Bi-modal

6. What is the cumulative relative frequency (starting from the bottom) of a score of 11?

    a. 27.45%   c. 94.31%

    b. 88.73%   d. 74.85%

7. In a study of post-traumatic stress, you obtain ratings of the number of flashbacks a person has in a month. If you have a sample of 75 individuals, and 40 have indicated they have had flashbacks twice in the last month, what is the relative frequency of those with two flashbacks in the last month?

    a. .63   c. .65

    b. .40   d. .53

Multiple-Choice Answers

1. C   5. B

2. B   6. B

3. A   7. D

4. D

Module Quiz

1. Using the following data, create a frequency table, cumulative frequency table, and relative frequency table.

7	8	4
10	6	4
5	7	7

27

3	5	6
4	3	9

2. You are investigating a measure of well-being in a geriatric population. The scale has a range of 0-50. The cumulative relative frequency for a score of 44 is 34% and the relative frequency is 4% for that interval. Why is the percentile rank of those with a score of 44 equal to 32% and not 34%?

3. You had a percentile rank of 73 in your high school class. What percentage of students were ranked above you?

4. You are trying to arrange the SAT scores of your high school in a frequency table. You realize that you need to use a grouped frequency table because of the large scale (0-1600), so you decide to create intervals of 400. Was this a good idea?

5. Name a variable that you think would have a) a negative distribution, b) a positive distribution, c) a bi-modal distribution, d) a normal distribution.

Quiz Answers

1.

X	f	Cum f	Rel f
10	1	15	6.67%
9	1	14	6.67%
8	1	13	6.67%
7	3	10	20.00%
6	2	8	13.33%
5	2	6	13.33%
4	3	3	20.00%
3	2	2	13.33%
2	0	0	0.00%
1	0	0	0.00%

2. Percentile ranks are based on the real limits of a score. The cumulative relative frequency was 34% with a relative frequency of 4%, indicating that 30% of the scores were below this

interval and 4% of the scores in this interval. Also, you can assume that the scores were evenly distributed within this interval. Thus, you have to add the amount of scores below (30%) to 1/2 of the amount of scores within it (4%), or 2%. This provides you with a percentile rank of 32%.

3. 27.

4. This was probably not a good idea. You would have 4 groups, and you would expect not many people to fall in the lowest group. Also, you would lose a great deal of precision, as most people would score between 800 and 1200, which would be just one interval.

5. Answers will vary.

Module 4

Learning Objectives

- Distinguish between normally distributed and nonnormally distributed data

- Select the best type of graph for data of a given scale

- Construct various types of graphs from data

- Apply graphing conventions so as not to distort the data

Module Summary

- Although frequency tables provide a neat method for organizing data, graphs can be an even more effective method for presenting information. Information that can be obtained from a graph includes the dispersion, clustering, and location of the majority of scores.

- *Stem and leaf displays* are similar to frequency tables but also provide a visual depiction of the frequency of each score. They are created by listing all of the data from highest to lowest values. The left column is the first digit of a score, and the right column contains every subsequent digit of all scores that start with the first digit. For very large data sets, you can represent the first digit twice in the left column. The right column should then be divided in half as well so that the top number displays scores that with a second digit of 5 or larger and the bottom number displays scores with a second digit less than 5.

- When creating a graph, the traditional rules are that the X-axis (abscissa) represents the intervals of the measured variable and the Y-axis (ordinate) represents the frequency of scores at each interval. In situations where there is a large frequency of cases for a particular score interval, you can divide the interval on the X-axis.

- There are some rules for creating a graph, which are as follows. (1) The Y-axis should be 3/4 the size of the X-axis. (2) With large data sets, you can collapse intervals on the X-axis so that there are at least 5 intervals but not more than 12 intervals. (3) Each interval on the *X*-axis must be equal to the others. (4) The *Y*-axis must be continuous. (5) The axes should not stretch or compress the data.

- *Histograms* are graphs for continuous data that indicate the frequency of a particular score by bars. The bars touch one another to indicate that a score could fall between the intervals on the *X*-axis.

- *Frequency curves* are alternatives to the histogram. A frequency curve is drawn by first creating a histogram and then connecting the midpoints of the adjacent bars with solid lines. This provides you with a visual representation of how your data are distributed, as it allows you to determine easily whether the data are clustered around a specific score or if they are spread out, and if they appear heavily lopsided or symmetrical.

- Frequency curves can take on multiple shapes. One of these shapes is the *normal curve*, which appears bell-shaped. This means that it is *symmetrical* in shape, with approximately half the scores falling above the peak (middle score) and half below. The other shapes involve *skew*, which refers to a lopsided distribution. Skew is produced by having a greater concentration of scores at either the upper or lower end of the distribution. Skew is named by the direction of the long tail. If there are a lot of high scores, the distribution is said to be *negatively skewed*, as there is a long tail on the left. If there are a lot of low scores, the distribution is said to be *positively skewed*, as there is a long tail on the right.

- *Kurtosis* refers to the amount of scores that are in the middle of the distribution. Distributions with a lot of scores in the center are referred to as *leptokurtic*. Alternatively, distributions with few scores in the center are referred to as *platykurtic*.

31

- The other two general shapes of the frequency curve are bi-modal and rectangular. *Bi-modal* distributions have more than one peak. *Rectangular* distributions are uniform in the spread of responses, which means the frequency of all the scores is the same.

- When graphing nominal data, it is more appropriate to use a bar graph or a pie graph, as these methods express the discrete nature of nominal variables. *Bar graphs* appear similar to histograms except that the bars are separated to indicate that a score could not fall between adjacent categories. *Pie graphs* place the nominal data in a circle, with slices to represent the different categories. The size of each slice indicates the amount of participants or cases in that particular category.

Computational Exercises

1. Create a stem and leaf display for these data: 47, 48, 48, 49, 49, 49 50, 50, 50 51, 51, 52.

2. Create a histogram that accurately portrays these data using 1-point intervals.

3. Create a frequency curve for this data using 1-point intervals.

4. Describe the shape of this distribution in terms of skewness and kurtosis.

5. If you were to create a positively skewed distribution, where would you need to add scores? A negatively skewed distribution? What would you need to do to make the distribution bi-modal? Rectangular?

6. State how you would expect the distributions for the following variables to appear:

    a. Time it takes a random group of people to run a mile

    b. Amount of money earned in the first year after completing college

    c. Age of football players

    d. Happiness of guests during a wedding

7. Which would be the appropriate methods for graphing the following data?

    a. Number of trophies won by tennis players

    b. Length of time in therapy for obsessive-compulsive disorder

    c. Number of words spoken by 3-year-olds

    d. The profit margins of six different companies for last year

Computational Answers

    1.

Stem	Leaf
5	2
5	1 1
5	0 0 0
4	9 9 9
4	8 8
4	7

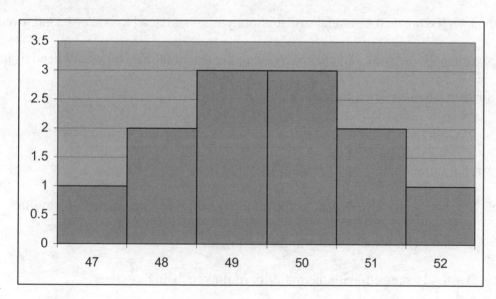

    2.

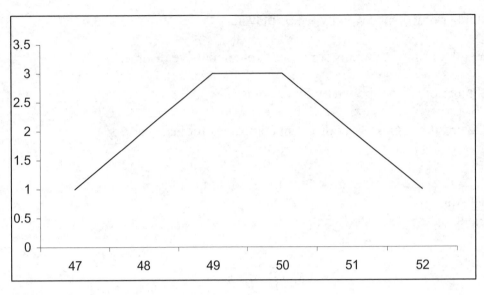

3.

4. The data are symmetrical, and there is no evidence of kurtosis.

5. To create a positively skewed distribution, you would need to add a large amount of scores below 47 or a few scores greater than 52. To create a negatively skewed distribution, you would need to add a large amount of scores above 52 or a few scores less than 47. To create a bi-modal distribution, you would need to add a second group of scores with a similar peak. To create a rectangular distribution, you would need to make it so all scores had an equal frequency.

6. a. Expect a positively skewed distribution, as there will be certain people who can run it quite quickly, but the majority will take a long time.

   b. Symmetrically distributed, as the majority will earn approximately the same amount, with fewer earning higher or lower.

   c. Positively skewed, with the majority of players being younger.

   d. Negatively skewed, with the majority of guests very happy.

7. a. Bar graph or pie chart.

   b. Histogram, frequency curve, stem and leaf.

34

c. Frequency curve, histogram.

d. Frequency curve, histogram.

True/False Questions

1. A score's percentile rank depends on the number of scores there are at that interval and the upper and lower limits of the score.

2. Pie charts and bar charts are excellent methods for graphing continuous data.

3. In a stem and leaf display, it can be useful to display each "stem" twice when you have a large amount of data.

4. A distribution with a lot of high scores and very few low scores would be considered positively skewed.

5. In a symmetrical distribution, the majority of the scores are above the midpoint.

6. Kurtosis refers to the height of the middle scores of a distribution.

True/False Answers

1. True	3. True	5. False
2. False	4. False	6. True

Short-Answer Questions

1. What can you do to better organize a stem and leaf display if you have a large amount of cases in a data set?

2. Why are the bars on a histogram connected?

3. What is the difference between a positively skewed distribution and a negatively skewed distribution?

4. You notice that the scores in a recent survey for how much viewers like a new TV show have a rectangular distribution. What does this mean about the responses in your sample?

5. What is kurtosis, and what are its two forms?

6. What do the slices on a pie graph represent?

Answers

1. You can list the stem value twice, with the upper listing corresponding to leaf values of 5 or greater and the lower listing corresponding to leaf values of 4 or less.

2. This indicates that the values could fall anywhere between the intervals.

3. A positively skewed distribution has a large amount of cases at the lower end of the scale and a few at the upper end. A negatively skewed distribution has a large amount of cases at the upper end of the scale and few toward the lower end.

4. Each score (interval) on your survey received an equal amount of responses.

5. Kurtosis refers to having either more or fewer scores in the center of a distribution, relative to a normal distribution. A distribution with more scores at the center is referred to as leptokurtic, and one with fewer is referred to as platykurtic.

6. The proportion of cases falling in each category. Pie charts are used for nominal data.

Multiple-Choice Questions

1. A professor announces to his class that a large portion of the class did really well on the last test. However, another large portion appeared to do poorly. How should the professor expect the distribution to look?

   a. Positively skewed   c. Bi-modal

   b. Negatively skewed   d. Symmetrical

2. The *X*-axis is sometimes referred to as the:

a. Abscissa

c. Platykurtic

b. Ordinate

d. Leptokurtic

3. In a study examining the rate at which a person can mentally rotate an object, the results indicate that the majority of people take 15 seconds, with an equal amount of participants falling above and below this time. You could expect the relative frequency curve of these scores to appear

a. Symmetric

c. Negatively skewed

b. Positively skewed

d. Bi-modal

4. What would the best method to graph data obtained for the question of "What is your favorite type of pie?"

a. Histogram

c. Bar graph

b. Relative frequency curve

d. Stem and leaf display

5. Bar and pie charts are best used for which type of variables?

a. Continuous

c. Nominal

b. Ordinal

d. Ratio

6. You administer a test of reading comprehension, with scores that can range from 1 to 20, to a group of 40 6-year-olds. Precisely two participants obtain each possible score. How will this distribution appear?

a. Rectangular

c. Positively skewed

b. Symmetrical

d. Negatively skewed

7. A common symptom of depression is a lack of desire to do things that are entertaining. If you were to ask a sample of 100 individuals with depression how many fun activities they have done in the past week, how would you expect the distribution to appear?

a. Rectangular

b. Symmetrical

c. Positively skewed

d. Negatively skewed

8. You are interested in determining how moviegoers rated a new film on opening day, on a scale of 1-8. After collecting data, you discover that most people in the sample provided ratings of either 4 or 5. How would you expect this distribution to appear?

a. Positively skewed

b. Negatively skewed

c. Leptokurtic

d. Platykurtic

Multiple-Choice Answers

1. C

2. A

3. A

4. C

5. C

6. A

7. C

8. C

Module Quiz

1. Create a histogram for the following data:

   1, 1, 1, 1, 2, 2, 2, 2, 2, 3, 3, 3, 3, 4, 4, 4, 5, 5, 6, 6, 7, 8, 9.

   What shape best describes this distribution?

2. You are working at an advertising firm and want to determine the interest of a focus group in a new product. If the scale ranges from 0 (no interest) to 10 (great interest), what type of distribution would be most preferable?

3. After giving his class a quiz worth 10 points (1-10 scale), your professor notices that the distribution was positively skewed. What conclusion should the professor draw from this information?

4. What types of variables are histograms best used for?

5. What does the height of a bar in a histogram represent?

Quiz Answers

1.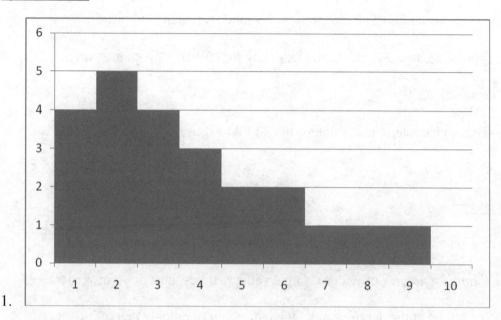

Positively skewed.

2. Negatively skewed.

3. That the test was very hard, and he might need to reteach the topics.

4. Continuous variables.

5. The frequency of scores in that particular grouping or category.

Module 5

Learning Objectives

- Calculate various measures of central tendency—mode, median, and mean

- Select the appropriate measure of central tendency for data of a given measurement scale and distribution shape

- Know the special characteristics of the mean that make it useful for further statistical calculations

Module Summary

- An entire data set can be described in a single number with a measure of central tendency. **Central tendency** provides a single value that best describes, or is most representative of, the entire set of scores. It enables you to quickly determine the center, which is usually the location of the majority of scores, of a large group of data.

- The *mode* is the most commonly occurring score in the data set. Although the mode is referred to as a measure of central tendency, it does not necessarily occur at the center of the data set (there may not be equal amounts of scores above and below the mode). The mode is the least stable measure of central tendency, meaning that it may change drastically from sample to sample of a population. This aspect of the mode reduces how often it is used.

- The *median* is the center score; half of the scores in the distribution are above the median, and half the scores in the distribution are below it. In other words, the median is the score that occurs at the 50th percentile. The median does not have to be an actual score. For example, the mean of the distribution 3, 4, 6, 7 is 5. If you do not know all of the specific scores in a data set, you can use the following formula for a precise measure of the median:

$$Median = LL + (i)\left(\frac{.5n - cum\,f\ below}{f}\right)$$

- The **mean** is the average score for a data set and is symbolized as M for samples and as μ for populations. The mean is the most commonly used and most stable measure of central tendency. The formula for a mean is as follows:

$$M = \frac{\sum X}{N}$$

- There are three important aspects to the mean. First, the numerical weight of scores above the mean is equal to the numerical weight of the scores below the mean. This indicates that if you were to find the distance of all the scores from the mean (i.e., $X - M$), the sum of the distances for the scores below the mean would be equal to the sum of the scores for the distances above the mean. Second, the mean includes all values of the data in its calculation, which indicates that each score in the distribution matters. Finally, the mean is also a sensitive measure of central tendency, in that a change in any score in the data set will change the mean.

- An extreme score in a data set, one that is drastically different from the others, is called an **outlier**. Outliers can influence which measure of central tendency is most appropriate to use. Because the mean is so sensitive to score values, the median may be a more appropriate measure if there are many outliers in a data set.

- The skew of a distribution will impact the location of the measures of central tendency. In a symmetrical distribution, the mean, median, and mode are all equal. In a skewed distribution, however, the values for the measures of central tendency are

41

as follows. Positively skewed distributions: Mode < Median < Mean. Negatively

skewed distributions: Mode > Median > Mean.

- It is not appropriate to report any single measure of central tendency when you have

    multimodal data. If reporting the mode, report multiple modes. A graph is the best

    method for displaying this distribution.

Computational Exercises

The following are the ratings of 23 newlyweds on a marital satisfaction inventory. The

scale ranges from 1 to 10. Use these data for questions 1-7.

6, 2, 3, 4, 5, 2, 6, 3, 4, 5, 6, 8, 7, 7, 5, 6, 8, 7, 5, 3, 4, 4, 5

1. What is the mean amount of marital satisfaction? What is the mode amount of marital

    satisfaction?

2. What is the median amount of marital satisfaction?

3. What is the numerical distance of the scores above the mean? What is the numerical

    distance of the scores below the mean? Which is greater?

4. Based on the measures of central tendency (use a ballpark estimate for the median),

    how would expect the distribution of these scores to appear? (Do not sketch your

    answer.)

5. If we were to add a score of 45 to this distribution, what would the new measures of

    central tendency be? Which measure would be the most influenced?

6. If we were to add another score of 5 to the original distribution, instead of the score of

    45, what would the new measures of central tendency be? Which score would be the

    most influenced?

7. In question 4, we added a score of 45 to the distribution. How would we expect the shape of the distribution to change if this score were to be added? How would we expect the shape to look if we were to add another score of 5?

The following are data obtained from tourists rating their experience at a particular hotel during a one-week stay in Hawaii. The scale that was used ranged from 1 to 50. The M = 29. Use the following data to answer question 8-11.

Score	Frequency
50	9
40	15
30	12
20	8
10	6

8. What is the modal response?

9. What is the median response?

10. How would you expect the distribution to appear, based on the measures of central tendency? (Do not sketch the distribution.)

11. Which measure of central tendency would best describe the distribution?

12. In a study looking at the distance people commute to work, the average distance was 40 miles. If the sum of all the distances in the sample was 200, what was n for the study?

Computational Answers

1. Mean = 5. Mode = 5.

2. $Median = LL + (i)\left(\dfrac{.5n - \text{cum f below}}{f}\right) = 4.5 + (1)\left(\dfrac{.5(23) - 9}{5}\right) = 5$

3. Sum above = 16. Sum below = −16. The difference is 0. Neither is greater. The mean is the balance point of the distribution.

4. It would appear normally distributed because the mean, median, and mode are equal.

43

5. Mean = 6.67, Median = 6.1, Mode = 5. The mean has changed the most.

6. Mean = 5, Median = 5, Mode = 5. None would have changed.

7. If we were to add a score of 45, the distribution would have a strong positive skew. If were to add another score of 5, the distribution would still appear normal (symmetrical).

8. 40.

9. $25 + (10)\left(\dfrac{.5(50) - 14}{12}\right) = 34.17.$

10. Positively skewed.

11. The median, because it is a skewed distribution.

12. $n = 5$.

True/False Questions

1. A measure of central tendency provides a single value that is considered highly representative of a distribution.

2. The mean is always the optimal measure of central tendency.

3. Each measure of central tendency (mean, median, and mode) can be found in the exact center of all distributions.

4. The mode is a highly stable measure of central tendency.

5. The numbers of scores falling above and below the median are always equal.

6. The median is the score that occurs at the 50th percentile rank.

7. One of the reasons that the mean is considered a strong measure of central tendency is that it includes every score in its computation.

8. The total distance of the scores below the mean will always be equal to the total distance of scores above the mean.

9. In a positively skewed distribution, you can expect all measures of central tendency to be equal.

10. There is no best single measure of central tendency to describe a bi-modal distribution.

11. Outliers are scores close to the mean.

12. Multiple outliers of different values will heavily influence the mode.

13. The median is considered to be a better measure of central tendency in a skewed distribution.

14. When describing a distribution with bi-modal data, you should report both modes.

15. In a negatively distribution you can expect the Mean > Median > Mode.

True/False Answers

1. True	6. True	11. False
2. False	7. True	12. False
3. False	8. True	13. True
4. False	9. False	14. True
5. True	10. True	15. False

Short-Answer Questions

1. Why are measures of central tendency useful?

2. What does it mean that a score is "representative of an entire data set"?

3. What is the mode? Is the mode a very stable measure of central tendency? Why or why not?

4. Define the median. How sensitive is the median to changes in the data set?

5. How is the mean different from the median and mode as a measure of central tendency?

6. Sometimes the mean is said to be the center of a see-saw. What does this mean?

7. What about the mean makes it the measure of central tendency that is the most sensitive to change?

8. What is an outlier? Describe a situation in which an outlier would have a heavy influence on the mean. Which measure of central tendency is most appropriate to use in this situation?

9. Compare and contrast the locations of the mean, median, and mode in the following distributions: positively skewed, negatively skewed, and symmetrical.

10. You have just completed a study and discover that you have a skewed distribution. Explain which measure of central tendency is the most appropriate to use with skewed data.

11. You are concerned that children are watching an excessive amount of violent TV. You collect a sample of viewing times and notice that the distribution is negatively skewed. If you wanted to report the largest number as a measure of central tendency, so as to make the biggest impact in the news, which measure of central tendency should you report?

12. A cupcake company has asked you to summarize its customers' opinions on a new type of cupcake filling. The company has provided an opinion scale with ratings of 1-5. If the data are normally distributed, which measure of central tendency should you report?

13. In looking over the data from the previous study, you notice that the data are symmetrical but leptokurtic. How would this affect the mean, median, and mode? (You may want to sketch the curve for help.)

Answers

1. They provide a single score to represent an entire set of numbers. They provide a quick "snapshot" of where the majority of the scores are located.

2. This indicates that the single score allows you to have a better understanding of where the majority of the scores from a data set are located.

3. The mode is the most commonly occurring score. It is not a very stable measure of central tendency because the addition of a few scores can drastically change the location of the mode. For example, a data set may have a mode of 3. However, after adding a few more scores with the value of 15, the mode may become 15.

4. The median is the center score, or the score that occurs at the 50th percentile. It is not very sensitive to changes in a data set. In large data sets, a large amount of scores would need to be added to heavily influence the median.

5. The mean differs from the median and mode in that it includes every score in its calculation. This makes it far more sensitive to changes in the data set than the other measures of central tendency.

6. This indicates that the total distance from the mean of scores above the mean is equal to the total distance from the mean of scores below the mean. This indicates that if you were to subtract the mean from each of the scores $(X - M)$, the absolute value of the sum of these differences for the scores above the mean will be equal to the absolute value of the sum of these differences for the scores below the mean.

7. The mean includes all of the scores in the distribution in its calculation. If you were to add any score to a data set (other than a score that is the mean itself), the mean would change, whereas this may not be the case with other measures of central tendency.

8. An outlier is an extreme score, or a score that differs greatly from all other scores. An example would be assessing the batting average of an entire baseball team. A very good player could drastically improve the average, whereas a very poor player could lower the average. The median would be a more appropriate measure of central tendency.

9. In a positively skewed distribution, you can expect the mode to be the smallest value, followed by the median, and the mean would be the largest value. In a negatively skewed distribution, the mean would be the smallest value, followed by the median, and finally the mode. In a symmetrical distribution, all three measures are equal.

10. The median would be most appropriate because the mean is being pulled in the direction of the tail. The mode "ignores" the fact that the distribution is skewed by reporting the most commonly occurring score. The median, however, falls between these scores and so is the most appropriate.

11. You should report the mode, because that will be the largest value.

12. You could report any measure, because all three would be equal.

13. This should have no influence on the measures of central tendency because the data are normally distributed.

Multiple-Choice Questions

1. You are interested in using the most *stable* measure of central tendency. Which measure should you use?

    a. Mean

    b. Median

    c. Mode

    d. All are equally stable

2. How should report central tendency when you have a bi-modal distribution?

    a. Report the mean

    b. Report one mode

    c. Report the median

    d. Report two modes

3. How many scores should you expect to fall above the mode?

    a. 50% of the scores

    b. The mode should be the center, such that the numerical distance of values above the mode is equal to the numerical distance of scores below the mode

    c. This would depend on the skew of the distribution

    d. There is no set amount of scores you should expect to fall above the mode

4. In the following set of scores, what is the median? 4, 5, 1, 3, 5, 3, 2, 6, 1, 9, 0, 9

    a. 2

    b. 2.5

    c. 3.5

    d. 4

5. In the following set of scores, what is the median? 3, 1, 3, 2, 4, 5, 9, 7

    a. 2

    b. 3

    c. 3.5

    d. 4

6. What is the difference between M and μ?

    a. M is for population means, and μ is for sample means

b. *M* is for sample means, and μ is for population means

c. *M* and μ have the same definition

d. *M* is for median, and μ is for mean

7. A high school football team is having tryouts and asks 5 participants to throw a football. Here is how far each person threw the ball (in feet): 79, 84, 93, 166, 88. Which score would be considered an outlier?

a. 79

c. 93

b. 84

d. 166

8. Using the information from question 7, the mean of the distances with the outlier would be ____, whereas the mean of the distances without the outlier would be ____.

a. 102 and 86

c. 86 and 88

b. 86 and 102

d. 88 and 86

9. You are collecting data for a study on hospital service. You have obtained data from patients at 30 different hospitals that have completed a satisfaction survey. The measures of central tendency are as follows: Mean, = 50; Median = 50; Mode = 50. What can you tell about all of the scores you have collected?

a. There are numerous outliers

b. The scale ranges from 1 to 50

c. The scores are distributed in a symmetrical shape

d. All participants provided identical information

The following information is the record sales of a music label's 9 recording artists, in thousands. Use this information for questions 10-15.

6, 8, 15, 10, 23, 17, 23, 32, 19

10. What is the average number of records sold by a recording artist at this label?

    a. 16.5   c. 17

    b. 22   d. 18.9

11. What is the modal number of record sales by an artist?

    a. 17   c. 15

    b. 27   d. 23

12. What is the median number of record sales by an artist?

    a. 17   c. 34

    b. 19   d. 23

13. How would you expect this distribution to appear without sketching it?

    a. Symmetrical   c. Negatively skewed

    b. Positively skewed   d. Bi-modal

14. The owner of the record label doesn't like the shape of the distribution of recent sales. He would like a stronger negative skew. To adjust the distribution so that it were more negatively skewed, where must the owner add scores?

    a. Above the mean   d. Obtain a few extreme

    b. Below the mean   scores above the mean

    c. Precisely at the mean

15. Here is the commute time (hours) per day of 7 people who work at the same company. 4, 1, .5, .75, 2, .25, 1.5. What is the average amount of time it takes these people to commute?

    a. 2.1   c. 1.43

    b. 1.67   d. 1.54

16. Using the information from question 15, the person with the 4-hour commute begins

    to work for a different company and is no longer included in the calculation. What is

    the new mean amount of time it takes for the remaining people to commute?

    a. 1.45   c. 1.47

    b. 1.07   d. 1.0

17. (Use information from questions 15 and 16.) To replace the person who left in

    question 16, the company hires a new person who commutes only .1 hour to work

    each day (he lives very close). What is the new mean amount of time that it takes for

    the workers to commute?

    a. 1.17   c. .64

    b. 1.0   d. .87

18. Using the original set of scores (from question 15), which measure of central

    tendency would best describe the distribution?

    a. Mean   c. Mode

    b. Median   d. None

19. A farmer is interested in finding out what vegetable is preferred by elementary school

    children. He asks them to select from the following categories: Peas, Broccoli,

    Carrots, Onions, and Celery. What measure of central tendency should he use to

    summarize his data?

    a. Mean   c. Mode

    b. Median   d. None

Multiple-Choice Answers

1. A   11. D

2. D   12. A

3. D   13. C

4. C   14. A

5. C   15. C

6. B   16. D

7. D   17. D

8. A   18. B

9. C   19. D

10. C

Module Quiz

Here are the results of a pop quiz given in an American history course. The quiz was

graded on a scale of 1-10.

$$2, 1, 9, 3, 2, 4, 5, 6, 7, 8, 2, 3, 7, 7, 2, 4, 7, 8, 5$$

1. What are the mean, median, and mode of these scores?

2. Which measure of central tendency would best describe this distribution?

3. You have just completed teaching a computer science course for the first time. Many

   of your students did not do well, and the distribution of final grades was positively

   skewed. You would rather not terrify your students at the start of your next class and

   decide to be a little deceptive with how you will report the performance of previous

   students. Which measure of central tendency would be most appropriate?

4. Following the example from 3, you only have 10 students enroll in your course the following semester. After the first test, you obtain the following grades (on a 0-70 scale): 45, 64, 57, 34, 65, 49, 44, 58, 58, 62. What is the mean of your class?

5. If you were to add one outlier to a symmetrical distribution, what would you expect to happen with regard to measures of central tendency?

Quiz Answers

1. Mean = 4.84; Median = $4.5 + (1)\left(\dfrac{.5(19) - 9}{2}\right) = 4.75$; Mode = 2 and 7 (both 4 occurrences).

2. The distribution is bi-modal, so there is no appropriate measure of central tendency, although both 2 and 7 may be reported.

3. You should report the mean because it is the highest value. (And you should make your tests easier!)

4. The mean would be 53.6.

5. The mean would change toward the direction of the outlier, but the mode and median would be relatively unaffected.

Module 6

Learning Objectives

- Calculate various measures of dispersion—range, variance, and standard deviation

- Select the appropriate measure of dispersion for data of a given distribution shape and for a given purpose

- Know the special characteristics of the standard deviation that make it useful for further statistical calculations

- Distinguish between descriptive and inferential formulas for the variance and standard deviation

Module Summary

- **Dispersion** is the extent that scores in a distribution are spread out or clustered together. Similar to measures of central tendency, measures of dispersion, or variability, are expressed as single values.

- The **range** is the difference between the highest score and the lowest score in a data set. The range is the simplest measure of dispersion and is very insensitive to changes in the distribution; adding scores to the center of the distribution will not affect the range. The only way the range can be impacted is by changing the most extreme scores. Due to these properties, the range is not often used.

- **Variance** (s^2 for samples; σ^2 for populations) is the average squared distance of each score from the mean. The formula for the variance is (depending on your instructor's preference):

$$s^2 = \frac{\sum (X - M)^2}{N} \text{ or } s^2 = \frac{\sum (X - M)^2}{n-1}$$

- The definition of the variance becomes more understandable by reviewing the formula. The numerator states that for every score in the distribution, you should obtain the ***deviation score***—the distance of each score from the mean. This is found by subtracting each score from the mean. All of the deviation scores sum to zero (this is a good way to check your work when finding variances). In order to proceed with obtaining the variance, you must square each deviation score. This will remove any signs (+/−) from the deviation scores and allow them to sum to a value other than zero. You then divide the sum of the squared deviation scores by the amount of scores in the sample (or by 1 fewer) to obtain the variance, or the ***average squared distance from the mean***.

- Unfortunately, the average squared distance from the mean is difficult to interpret. The variance tells you the average distance from the mean in area units (which we are unable to interpret) as opposed to linear units (which we normally use). ***Linear*** distance is distance in original units, or regular score points (i.e., the linear distance between 3 and 5 is 2). To revert the variance (in area units) back to linear units, you take the square root of the variance. This result is referred to as the standard deviation. The ***standard deviation*** (s for samples; σ for populations) is the average *linear* distance of scores from the mean. The formula for standard deviation is (depending on the formula used for the variance, from above):

$$ s \; = \sqrt{\frac{\sum (X - M)^2}{N}} \quad \text{or} \; s \; = \sqrt{\frac{\sum (X - M)^2}{n-1}} $$

- The standard deviation is a *standardized* measure of dispersion, indicating that it can be used when working with a specific type of distribution called the normal curve. The normal curve will be discussed at length in later chapters.

- Not all measures of linear dispersion are standardized. One unstandardized measure of dispersion is the ***average absolute deviation***, which uses the absolute value of each deviation score rather than the squaring technique. Although this provides a more intuitive measure of dispersion, the measure does not fall within known locations on the normal curve, which limits its use.

- A statistic (or parameter) that is used to measure dispersion in describing a sample is called a ***descriptive statistic***. Alternatively, you can use sample statistics to make guesses about larger populations. This is commonly referred to as ***inferential statistics***. You are using the sample data to make inferences (guesses) about the larger population.

- When using sample variances and standard deviations to infer about a larger population, it is important to note that your estimate will not be precise. In other words, your sample variance likely will not equal the population variance. In fact, you can be almost certain that the sample variance will be less than the population variance. Placing $n - 1$ in the denominator of the variance and standard deviation formulas will adjust for this bias when estimating the population standard deviation from a sample. This adjustment will increase the variance (or standard deviation), which will help to better approximate the population variance (or standard deviation). Because of the inferential use of the standard deviation and variance that will be introduced later, some instructors prefer to use $n - 1$ in the denominator of even the descriptive standard deviation and variance. Ask your instructor which formula is preferred.

Computational Exercises

The following are test grades for students in your European History class after the first test.

87	89	99	99
98	93	67	89
90	93	70	87
65	94	82	95
70	79	66	90

1. Find the deviation score for each test grade. What is the sum of these deviations?

2. Find the variance for the grades. Find the standard deviation for the grades.

3. How many standard deviations from the mean is the person with the highest grade? The person with the lowest score?

4. One of those who received a 99 is an exceptional student and turned in an extra-credit project that was worth an additional 10 points on the test. What would the variance and standard deviation be with this person's correct grade? (Recalculate the measures of dispersion using this new high score.)

5. Reflecting back on your response to question 4, which measure of dispersion had the largest change in absolute units?

6. How many of the original grades are between 1 and 2 standard deviations above the mean?

The manager of a shoe store is interested in determining how many of each shoe size was sold in the past day. The store has made 10 sales, of shoes with the following sizes.

7	12	6
11	8	12
13	8	
7	6	

7. Find the variance and standard deviation for these shoe sizes.

8. At the last few minutes of the store's business hours , three people run in, stating that they are in a shoe emergency, and ask to purchase shoes. The sizes of these three new customers are 8, 6, and 10. If they each purchase a pair of shoes, what will be the new standard deviation?

9. How many of these 13 (including those added in question 8) people fell within one standard deviation of the mean?

10. What is the average absolute deviation of the original shoe sizes? How does this compare with the standard deviation?

11. If you were to estimate the population variance of shoe sizes of people who shop at this store from this sample, what formula would you use (using the original 10 scores)? What would be the population variance estimated from this sample?

12. The store manager collects similar data for the following day and finds the mean is substantially higher, but the standard deviation is now 7.9. Which mean is more representative of the average shoe size of those who shop at this store?

Computational Answers

1.

X	X– M
87	1.9
98	12.9
90	4.9
65	–20.1
70	–15.1
89	3.9
93	7.9
93	7.9
94	8.9
79	–6.1
99	13.9
67	–18.1

70	−15.1
82	−3.1
66	−19.1
99	13.9
89	3.9
87	1.9
95	9.9
90	4.9

2. The sum of the deviations is 0.00. Variance = 126.99 (using *N*) OR = 133.67 (using *n* − 1); standard deviation = 11.27 (using *N*) OR = 11.56 (using *n* − 1).

3. The highest grade is approximately 1.20 standard deviations above the mean. The lowest grade is approximately 1.7 standard deviations below the mean.

4. The new variance = 145.64 (using *N*) OR = 153.31 (using *n* − 1). The new standard deviation = 12.07 (using *N*) OR = 123.38 (using *n* − 1).

5. The variance is most affected by this change.

6. Three scores: 98, 99, and 99.

7. The variance = 6.60 (using *N*) OR = 7.33 (using *n* − 1). The standard deviation = 2.57 (using *N*) OR = 2.71 (using *n* − 1).

8. The new standard deviation would be 2.42 (using *N*) OR = 2.52 (using *n* − 1).

9. There are 7 scores that fall within one standard deviation above and below the mean.

10. Average absolute deviation = 2.18. It is slightly smaller than the standard deviation.

11. You would use the formula with n-1 in the denominator because the sample is less than 30. The estimated population variance would be 7.33.

$$s^2 = \frac{\sum (X - M)^2}{n - 1}$$

12. The first mean is more representative because the scores are more tightly clustered about the mean.

True/False Questions

1. The range is a very sensitive measure of central tendency.

2. Adding 10 different scores to the center of a data set will not impact the range.

3. A deviation score provides a measure of the score's distance from the mean.

4. The sign (+/−) of a deviation score indicates its location in relationship to the mean.

5. Deviation scores always sum to 1.

6. The variance is the square of the average distance of scores from the mean, measured in squared distance units.

7. Standard deviations from different samples of the same population can be compared.

8. The standard deviation is found from the square root of the variance in order to revert the measure to linear units.

9. The standard deviation is the most commonly used measure of dispersion because of its interpretability and applicability to the normal curve.

10. The average absolute deviation is used more frequently than the standard deviation.

11. Descriptive statistics are used to describe the characteristics of a sample.

12. The mean and standard deviation are examples of inferential statistics.

13. When using a sample to infer about a population, you should use $n - 1$ in the denominator.

14. Using $n - 1$ in the denominator corrects for the bias of sample variance being larger than the population variance.

15. There is a debate in the social sciences regarding the appropriate use of N as opposed to $n - 1$ in the denominator of sample variances.

1.	False	6.	True	11.	True
2.	True	7.	True	12.	False
3.	True	8.	True	13.	True
4.	True	9.	True	14.	False
5.	False	10.	False	15.	True

Short-Answer Questions

1. Why are measures of dispersion important when describing a sample?

2. What aspects of the range make it a poor measure of dispersion?

3. What are the steps involved in finding the variance?

4. Why is it necessary to square the deviation scores when finding the variance?

5. What unit of measurement is used for the variance? What unit of measurement is used for the standard deviation? How are these different?

6. How are area units changed to linear units?

7. Why can we always expect the deviation scores to sum to zero?

8. In a high school gym class, the teacher is interested in assessing the average and standard deviation of the amount of time it takes the students run a mile. The teacher notices that one student is exceptionally quick and completes the mile in a time that is 4 standard deviations below the mean. Would you consider this person an outlier? Why or why not?

9. How would all three measures of dispersion (variance, range, and standard deviation) be impacted by adding a large amount of scores close to the mean of any given distribution?

10. What does it mean to be a standardized measure?

11. What is the difference between a standard deviation and an average absolute deviation?

12. Why is the average absolute deviation not commonly used?

Answers

1. Dispersion is important because it indicates the extent that scores cluster around the mean or are very distant from the mean. This will help you to determine how viable the mean is as a single descriptor of the data set. For example, a 0-10 scale with a mean of 5 and a standard deviation of 1 indicates that on average, a score will deviate 1 point from the mean. This suggests that the majority of scores will fall between 4 and 6, making the mean a very good descriptor. However, if the mean was 5 and the standard deviation was 5, that means the average deviation from the mean was 5 points, indicating that the scores fall everywhere on the scale.

2. The range does not consider all of the scores in the distribution. Also, it can be heavily influenced by extreme scores (drastically different upper or lower scores), and it is unaffected by the addition of non-extreme scores (scores that are not the upper or lower limits).

3. First, the deviation scores must be found. Then these scores must be squared. The squared deviation scores are then summed. Finally, this summed squared deviation score is divided by N or by $n - 1$.

4. This is because the deviation scores will sum to 0, which would indicate that there is no variability in the sample. Squaring the deviation scores creates a positive sum.

5. Variance is in area units. The standard deviation is in linear units. The difference is that the area units are squared, meaning they are not able to be applied directly to the original scale of measurement. The linear units are in the same metric as the original scale.

6. Area units are changed to linear units by taking the square root of the area unit.

7. Deviation scores always sum to zero because they represent each score's numerical distance from the mean. Because the mean is the numerical center of the distribution, the distance of scores above the mean will always equal the distance below the mean.

8. This person would be considered an outlier because he or she fell so far away from the mean.

9. The range would not be impacted at all by adding scores to the center of the distribution. However, the variance and standard deviation would become smaller.

10. This means that the values all fall on the same scale. The standard deviation is on the scale of the normal curve, which has unique properties that will be encountered in later sections.

11. A standard deviation represents the standardized average deviation which is applicable to the normal curve. The average absolute deviation uses the absolute value of the deviation scores.

12. The average absolute deviation does not fall within known places on the normal curve, which is extremely useful in the practice of statistics. Because of this, the absolute average deviation is not commonly used.

Multiple-Choice Questions

The coach of a basketball team is interested in assessing how well his team shoots free throws. Here is the number of successful free throws that each member of his team made during their last practice session. Use this information for question 1-5.

10	0
3	10
11	3
1	1
0	1
7	8

1. The coach is asked to provide a quick measure of the dispersion for his team's free throws. What is the range for his team?

    a. 0

    b. 1

    c. 11

    d. 12

2. The coach has more time on his hands and determines the variance. What is it?

    a. 11.29 (using N), 12.32 (using $n - 1$)

    b. 15.52 (using N), 16.92 (using $n - 1$)

    c. 12.11 (using N), 13.21 (using $n - 1$)

    d. 4.18 (using N), 4.56 (using $n - 1$)

3. The coach realizes that the variance is difficult to interpret and now wants to know the standard deviation for his team. What is the team's standard deviation?

    a. 5.09 (using N), 5.32 (using $n - 1$)

    b. 1.18 (using N),1.23 (using $n - 1$)

    c. 15.52 (using N), (using $n - 1$)

    d. 3.94 (using N), 4.11 (using $n - 1$)

4. Shortly after obtaining these data, a new player is added to the team who makes 10 free throws. If you were to incorporate this new player's score, which measure of dispersion would not be affected?

    a. Variance

    b. Standard deviation

    c. Range

    d. Mean

65

5. The coach is interested in using the data from his team to learn about the amount of free throws made by all of the teams in the league. This is an example of

    a. Descriptive statistics

    b. Inferential statistics

    c. Central tendency

    d. Dispersion

6. A distribution of scores is discovered to be highly leptokurtic. How much dispersion would you expect?

    a. A small amount because many scores are close to the mean

    b. A large amount because many scores are close to the mean

    c. A small amount because many scores are distant from the mean

    d. A large amount because many scores are distant from the mean

7. How much dispersion can you expect with a constant?

    a. A great deal of variability

    b. A moderate amount of variability

    c. Minimal variability

    d. No variability

8. You want to find the standard deviation for a set of scores in order to infer the results to a larger group of scores. Which formula should you use?

    a. $s^2 = \dfrac{\sum(X - M)^2}{N}$

    c. $s^2 = \dfrac{\sum(X - M)^2}{n - 1}$

    b. $s = \sqrt{\dfrac{\sum(X - M)^2}{N}}$

    d. $s = \sqrt{\dfrac{\sum(X - M)^2}{n - 1}}$

Here are the responses of 10 individuals on a survey assessing satisfaction at the local Department of Motor Vehicles (scale is 1-10). Use this information for question 9-14.

5	3
5	6
7	6
1	1
1	6

9. What is the range of these ratings?

    a. 1   c. 6

    b. 7   d. 5

10. What is the variance of these ratings?

    a. 6.78 (using N), 7.40 (using $n-1$)   c. 7.89 (using N), 8.60 (using $n-1$)

    b. 5.09 (using N), 5.55 (using $n-1$)   d. 4.21 (using N), 4.59 (using $n-1$)

11. What is the standard deviation of these ratings?

    a. 2.26 (using N), 2.38 (using $n-1$)   c. 5.70 (using N), 5.95 (using $n-1$)

    b. 3.47(using N), 3.62 (using $n-1$)   d. 6.87 (using N), 7.18 (using $n-1$)

12. What is the average absolute deviation of these ratings?

    a. 2.08   c. 2.27

    b. 3.40   d. 4.58

13. How many scores are more than 1 standard deviation away from the mean?

    a. 1   c. 3

    b. 2   d. 4

14. What would the standard deviation be if the supervisor of the DMV wanted to make an estimate about the satisfaction of those that have used the DMV?

    a. 2.38

    b. 1.52

    c. 1.1

    d. 7.5

15. Late one night, Javier is working on some marketing data that are symmetrically distributed. He is very tired and accidentally uses the median instead of the mean when calculating his deviation scores. How will this affect his standard deviation?

    a. It will increase it

    b. It will decrease it

    c. It will not affect it

    d. You need more information

16. You are interested in studying the eye color of your fellow classmates. You obtain the following data: Blue, Blue, Brown, Brown, Brown, Brown, Green, Green, Gray. What would the standard deviation for this distribution?

    a. 1

    b. 2

    c. 3

    d. You can't calculate a standard deviation for nominal data

Mr. Jones is having difficulty sleeping. He decides to chart the number of hours he has slept over the past week. Use this information for question 17-19.

$$4, 3, 7, 8, 2$$

17. What is the variance of the number of hours he has slept?

    a. 48.85 (using N), 60.56 (using $n - 1$)

    b. 5.36 (using N), 6.70 (using $n - 1$)

    c. 7.45 (using N), 9.31 (using $n - 1$)

    d. 6.89 (using N), 8.61 (using $n - 1$)

18. What is the standard deviation of the amount of hours he has slept?

    a. 2.32 (using N), 2.59 (using $n - 1$)

    b. 3.45 (using N), 3.86 (using $n - 1$)

    c. 6.82 (using N), 7.62 (using $n - 1$)

    d. 7.41 (using N), 8.28 (using $n - 1$)

19. How many nights has Mr. Jones slept within one standard deviation of the mean?

     a. 1   c. 3

     b. 2   d. 4

Multiple-Choice Answers

1. C   11. A

2. B   12. A

3. D   13. D

4. C   14. A

5. B   15. C

6. A   16. D

7. D   17. B

8. D   18. A

9. C   19. C

10. B

Module Quiz

1. Researchers are interested in determining how many food pellets rats on a new diet drug have eaten. They discover the average amount of food eaten is 4.5 pellets and the standard deviation 0.75. What is the variance of food pellets?

2. Why do we use $n - 1$ in the denominator of variance when estimated a population variance from a sample?

3. In general, how would adding outliers to a distribution impact dispersion?

4. What are descriptive statistics? What are examples of this type of statistic?

5. What are inferential statistics? Why are they important in our study of statistics?

Quiz Answers

1. The variance is 0.56.

2. $n - 1$ is used in the denominator to correct for a sample's underestimation of the variability in a population.

3. It would increase dispersion.

4. Descriptive statistics are those that summarize the data in a sample. Examples of this type of statistic are measures of central tendency and dispersion.

5. Inferential statistics are used to make estimates about a population from a sample. This is important because it can be very difficult to directly measure a population.

Module 7

Learning Objectives

- Know the history of the normal distribution

- Know the conditions under which data will be distributed normally

- Know the number of standard deviations in a normal distribution

- Know the percentage of cases falling between whole standard deviation values in a normal distribution

Module Summary

- The normal curve is a symmetric, bell-shaped curve that has an ***inflection point*** (meaning the curve bends) at one standard deviation above and below the mean. The shape of the normal curve is how a distribution with an infinite number of scores would appear. This means that the normal curve is theoretical, as opposed to an actual distribution of scores. However, we can expect that scores will distribute themselves in a manner similar to that of the normal distribution, especially as the size of the sample grows beyond $n > 30$.

- The data sets of this book will rarely fall in a perfect the normal curve. However, the properties of the normal curve are ***robust*** to distributions that may violate its shape. This indicates that although our data may not look perfectly normal, you can still use the special features of the normal curve to understand our sample.

- The benefit of using the normal curve is that you are able to determine the proportion (percentage) of scores that will fall within one, two, and three standard deviations of the mean. In a normal distribution, you can always expect that about 68% of the scores will fall within (+/–) one standard deviation of the mean, that about 95% of the scores will fall within (+/–) two standard deviations of the mean, and that about 99% of the scores will fall within

71

(+/–) three standard deviations of the mean. This means that a score that falls 1 standard deviation above the mean will fall in the same place on the normal curve, regardless of what is being measured. Aside from knowing the proportion of scores at a specific standard deviation, you can use the normal curve to determine the exact proportion of the curve above and below any specific score.

Computational Exercises

1. The mean grade on a French test was 74 with a standard deviation of 6. If you scored 0.5 standard deviations above the mean, what was your grade on the test?

2. What percentage of the normal distribution is greater than 2 standard deviations away from the mean? Greater than 1 standard deviation away from the mean?

3. What percentage of the normal distribution is greater than the mean? What percentage of the normal distribution is less than two standard deviations below the mean? What percentage of the normal distribution is within one standard deviation above the mean?

4. Van has a score that is 2 standard deviations above the mean on a History test. What proportion of the distribution fell above his score?

5. 13.59% of scores in normal curve fall between +1 and +2 standard deviations above the mean. What percent of scores fall between –1 and –2 standard deviations below the mean?

Computational Answers

1. 77.

2. Approximately 5% of the distribution is greater than 2 standard deviations away from the mean. Approximately 32% of the normal distribution is greater than 1 standard deviation away from the mean.

3. 50% of the normal distribution is greater than the mean. Approximately 2.5% of the normal distribution is less than two standard deviations below the mean. Approximately 34% of the normal distribution is within one standard deviation above the mean.

4. 2.28% of the distribution fell above Van's score.

5. 13.59%.

True/False Questions

1. The majority of the scores of the normal curve occur within 1 standard deviation of the mean.

2. The end points of the normal curve represent scores with a frequency of zero.

3. You can still use the assumptions associated with the normal distribution with distributions whose shapes slightly deviate from normality.

4. Approximately 50% of the scores occur within 3 standard deviations of the mean.

5. The mean of IQ scores is 100 with a standard deviation of 15. The mean of the Minnesota Multiphasic Personality Inventory (MMPI) scores is 50, with a standard deviation of 10. An IQ score of 85 and an MMPI score of 40 fall in the same place on the normal curve.

True/False Answers

1. True   3. True   5. True

2. False   4. False

Short-Answer Questions

1. What is an inflection point? Why are they important in the normal curve?

2. What aspect of the normal curve makes it very useful in statistics?

3. In regards to statistics, what does it mean for something to be robust?

4. What does it mean for a score to fall at the 85% percentile of the normal curve?

5. If you were to find that you have a have score that is worse than 2.5% of the population, where would you fall in the normal distribution?

Answers

1. An inflection point is the place on the normal curve where the frequency curve drastically changes direction. On the normal curve, the first major inflection point represents one standard deviation above and below the mean.

2. The most useful aspect of the normal curve is that the exact proportion (percentage) of the curve above and below any given point is known. This tells us how many scores we can expect between pairs of standard deviations away from the mean. Also, it helps us to determine the amount of people above and below any particular score when the data are normally distributed.

3. It means that you can still use the statistical procedure even though some of the required assumptions of the procedure have been violated. For example, you can still use the rules of the normal curve with data that do not perfectly conform to the shape of the normal curve.

4. It means that 85% of the scores were either equal to or less than this particular score. Also, 15% of the scores were greater than this score.

5. You would fall two standard deviations above the mean, because this represents a z-score of 2.

Multiple-Choice Questions

1. Almost of all of the scores in a distribution ($\approx .99$) fall within how many standard deviations of the mean?

   a. 0   c. 2

   b. 1   d. 3

2. As you move further away from the mean in a normal distribution, the frequency of scores

a. Increases   c. Stays the same

b. Decreases   d. The answer depends on the data set

3. The majority of scores in a normal distribution fall where in relation to the inflection point?

    a. Outside the inflection points   c. Precisely at the inflection point

    b. In between the inflection points   d. The answer depends on the data

4. The population for the SAT has a mean = 500 and s = 100. A score of 300 has as many people below it as what score above it?

    a. 300   c. 600

    b. 500   d. 700

Multiple-Choice Answers

1. D   3. B

2. B   4. D

Module Quiz

1. What proportion of the population falls within 1 standard deviation above and below the mean? Two standard deviations above the mean?

2. The population mean for SAT scores has an M = 500 and an SD = 100. Between which two scores would ≈ 99% of the population fall between?

3. Harry's score on a math test is two standard deviations above Ron's. If Ron scored 1 standard deviation above the mean and the descriptive statistics for the class were M = 50, SD = 15, what were Ron's and Harry's raw scores?

4. Although Harry did better than Ron in math, Ron is a stronger English student. If the descriptive statistics for an English class test were M = 79, SD = 7, Ron obtained a score that was approximately 1.3 SDs above the mean, and Harry obtained a score that was approximately 0.12 SDs above the mean, what were their raw scores?

75

1. 68% fall between 1 standard deviation above and below the mean. 47.5% fall two standard deviations above the mean.

2. 99% of the population would fall between 3 *SD* above and below the mean, which corresponds to scores of 200 and 800.

3. Ron's score was 35, and Harry's was 65.

4. Ron's score was 88, and Harry's was 80.

Module 8

Learning Objectives

- Know the advantages of standard scores over raw scores

- Understand the process of rescaling numerator units into denominator units

- Calculate a z score

- Use a normal curve table to determine percentages above, below, or between given z scores

Module Summary

- **Standard scores** are scores that have been converted to a standard scale of measurement. This means that they can be compared to other scores that have been placed on the same scale. In other words, you can compare a person's height in feet to their weight in pounds by converting both values to standard scores. This is in contrast to raw scores, which are the scores on their original scale.

- Using **z scores** standardizes scores on the scale of standard deviation units. A z score of 1 means that the raw score fell one standard deviation above the mean. The sign of a z score (+/−) indicates the location of the score with regard to the mean (− below the mean, + above the mean). The formula for finding a z score is:

$$z = \frac{X - M}{s}$$

- The z score formula is merely a conversion. First you, determine how far your score is from the mean in raw units ($X - M$). Then, you determine how many standard deviations this distance is by dividing by s. This is similar to converting feet to inches or yards to feet.

- After converting a score to a z score, you are able to use it in conjunction with your knowledge of the normal curve because z scores are normally distributed. You can find the exact percent location of your z score in a normal distribution. In other words, you can find the percentage of scores above your z score and the percentage of scores below your z score. These percentages are found using a Normal Curve Table, which can be found in Appendix A of your text. The percentages on the Normal Curve Table indicate the percentage of scores at or below a particular z score value. To determine the percentage of scores above a z score value, subtract the percentage from 1. Thus, a z score of 1.33 will be greater than or equal to 90.82% scores of the distribution and less than 9.18% scores of the distribution. Finally, you can use the normal table to determine how many scores fall between a particular z score and the mean by subtracting 0.5 from the percentage found in the table.

- An important aspect of a z score is that it includes a distribution's central tendency (in the mean) and dispersion (standard deviation) in its calculation. This enables you to compare scores from completely different scales (such as tests in different classes) once they have been converted to z scores. Yes, this means you can finally compare apples to oranges!

Computational Exercises

1. Using a distribution of with a mean of 30 and a standard deviation of 4, find the z scores for the following scores:

   a. 28   c. 39

   b. 14   d. 4

2. Sana just took a very difficult cognitive psychology test and a very easy calculus test. On the psychology test, she earned a grade of 78 and found out the class had a mean of 72 with a standard deviation of 4. In contrast, she earned a grade of 87 on the calculus test and found out the class had a mean of 89 and a standard deviation of 6. How would you explain to Sana that she should feel good about her psychology test grade?

3. Here are the scores obtained by each of the 5 members of Tim's bowling team during last night's tournament. Convert each of the scores to *z* scores:

    133, 159, 112, 131, 169

4. Two of the local little league baseball teams want to have a contest to determine which team can catch more fly balls. Here are the means and standard deviations for amount of fly balls caught by all of the members of each team.

    Team A: Mean = 7.9; SD = 2.3

    Team B: Mean = 8.3; SD = 3.9

    Saul, Team A's best player, caught 10 fly balls. Ari, Team B's best player, also caught 10 fly balls. Which of these two players is better, relative to the performance of his team?

5. Here is the number of court cases a particular judge saw per day for the past week.

    7, 5, 9, 3, 4

    What is the *z* score for the day in which he saw the most cases? What percentage falls between this score and the mean?

Here are the scores obtained by each of the 5 members of Tim's bowling team during last night's tournament. Use this information for question 6 and 7.

    133
    159
    112
    131
    169

6. What percentage of the normal curve is equal to or below the person with the lowest score? What about the person with the highest score?

7. The person who bowled a 159 thinks that she is better than the average bowler. How much of a percentage above the mean is this person?

8. Two of the local little league baseball teams want to have a contest to determine which team can catch more fly balls. Here are the means and standard deviations for amount of fly balls caught by all of the members of each team.

Team A: Mean = 7.9; SD = 2.3

Team B: Mean = 8.3; SD = 3.9

Jose caught 4 fly balls and is on Team A. What proportion of the normal curve falls between Jose and the mean of Team A?

Computational Answers

1. a.–.5 b. –4 c. 2.25 d. –6.5

2. You should tell her that she did much better on the psychology test relative to everyone else as she had a z score of 1.5. However, on the calculus test, she didn't do as well compared to everyone else, as she had a z score of –.33.

3. $M = 140.8$, $s = 20.56$

X	z
133.00	–0.38
159.00	0.89
112.00	–1.40
131.00	–0.48
169.00	1.37

4. Saul's z score = .91. Ari's z score = .44. Saul is better relative to his team than Ari is.

5. The z score for day 3 of 9 was 1.58 ($M = 5.6$; $s = 2.15$). Approximately 44% of the normal curve falls between this score and the mean.

6. 8% (.08). 91% (.91).

7. 31% of the normal distribution falls between a z score of .89 and the mean of 0.

8. Approximately 46% of the normal curve falls between Jose and the mean of Team A.

True/False Questions

1. Standardized scores tell you the position of a score in reference to all of the other scores in a distribution.

2. A z score of 1.5 corresponds to a score that is 1.5 standard deviations above the mean.

3. A z score tells you the location of a score in relation to the mean.

4. The scale of z scores is the scale of variance units.

5. The proportion of the normal curve below a z score of .12 is .45.

6. You can compare scores from entirely different groups after converting the scores to z scores.

7. Brand A sells an average amount of 5.4 units per day with a standard deviation of 1.2. Brand B sells an average amount of 3.2 units per day with a standard deviation of 2.5. A z score of 1 for these two brands would represent different raw scores but indicate a score that fell in the same place on the normal curve.

True/False Answers

1. True	4. False	7. True
2. True	5. False	
3. True	6. True	

Short-Answer Questions

1. What two pieces of information can be found from a z score?

2. You want to compare the average amount of nail polish used by a nail salon in a week to the average amount of lettuce sold per week by a grocery store. What type of scores could you use to make this comparison? What about this type of score helps you to make this comparison?

3. How does the formula for a z score "rescale" raw scores?

4. What will always be the mean and standard deviation of a distribution that has been converted to z scores?

5. The students in Course A earn a mean grade of 85 with a standard deviation of 2.32 on the first test. The students in Course B earn a mean grade of 93 with a standard deviation of 3.84. You are enrolled in both courses and discover that your z score for Course A and B was 0.48. Does this mean that you have the same grade in both courses? Why or why not?

Answers

1. The distance of the score from the mean in standard deviation units and the direction of the score from the mean (either above or below).

2. You would need to use a standardized score. Standardized scores are on the same scale, so they can be directly compared with one another.

3. A z score tells you the location of a score in relation to the mean in standard deviation units. This is obtained by first subtracting the score from the mean (numerator). Then this deviation is scaled in standard deviation units by dividing by the distribution's standard deviation.

4. The mean will always be 0, and the standard deviation will always be 1.

5. The grades are not the same in both courses because the means and standard deviations of the distributions are different. However, it does mean that your score fell in the same location relative to everyone else in those classes.

Multiple-Choice Questions

1. What percentage of the normal curve falls between $z = 0.64$ and $z = 0.89$?

   a. 0.2389   c. 0.0744

   b. 0.3133   d. 0.9256

2. What percentage of the normal curve falls between $z = -0.93$ and $z = 1.60$?

a. 0.8238   c. 0.7690

b. 0.3238   d. 0.4452

3. What percentage of the normal curve falls outside of the area between $z = 1.5$ and $z = .34$?

    a. 0.4338   c. 0.9331

    b. 0.6331   d. 0.3001

4. If a distribution of scores has a mean of 15 and a standard deviation of 1.5, what is the z score

    for a score of 18?

    a. 0.5   c. 1.5

    b. 1   d. 2

5. The director of a new horror movie wants to know how scared her daughter was while

    watching the film. She asks her daughter to rate her fear on a scale of 1 to 15. If the average

    rating of fear for this movie is $M = 6$ with an $SD = 0.75$, where in the normal distribution

    would the director's daughter if she gave a rating of 5?

    a. Higher than 25.23%   c. Higher than 9.18%

    b. Lower than 78.65%   d. Lower than 36.47%

6. The local police department wants to determine the amount of respect that the local citizens

    have toward them. They obtain data from all of the residents and find that the $M = 3$ and the

    $SD = 2.7$. In looking through the data, they notice that one participant rated his respect level

    at a 9. What is the z score for this person?

    a. 1.22   c. 0.79

    b. 2.22   d. −1.62

7. What percentage of the normal curve is greater than a z of .34?

    a. 0.3669   b. 0.4521

c. 0.6331   d. 0.7892

8. What percentage of the normal curve falls between a $z = -1.2$ and the mean?

    a. 0.0214   c. 0.7842

    b. 0.3849   d. 0.6254

Multiple-Choice Answers

1. C   5. C

2. C   6. B

3. A   7. A

4. D   8. B

Module Quiz

1. The raw scores of a distribution have an $M = 78$ and an $SD = 12.3$. You obtain a z score of -2.3. What is your corresponding raw score?

2. Three electronics companies are trying to beat each other's prices on car stereos. Here is a table with the average original prices, standard deviations, and the current sale price of their models. Who is offering the best deal?

Company	Mean	SD	Sale
A	450	23	389
B	375	12	359
C	475	57	401

3. Two brothers are having an eating competition. Mike is seeing how many hot dogs he can eat in a minute, and Jeff is seeing how many hamburgers he can eat in a minute. If Mike eats 7 hot dogs and usually eats $M = 4.5$, $SD = 1$, and Jeff eats 6 hamburgers and usually eats $M = 3.2$, $SD = 2.1$, who ate more relative to their normal amount of consumption?

4. A makeup company is attempting to lower its prices in order to improve sales. It decides to charge $5 for all of its products. If eye liner has $M = \$6.50$, $SD = \$0.35$, and foundation has $M = \$7.45$, $SD = \$0.78$, which of these products has the bigger price reduction?

5. A car advertisement states that its new models are able to accelerate at a rate that is two standard deviations above the mean. If the average acceleration of a car (in miles per hour) is 50 with an $SD = 5.3$, what is the acceleration of the car being advertised?

6. Using the information provided in questions 3 and 4 of Module 7's quiz, who is the stronger student in their respective subject (Harry as a math student or Ron as an English student)? Why?

Quiz Answers

1. 49.71.

2. Company A.

3. Mike.

4. Eye liner.

5. 60.6.

6. Ron is a stronger in English than Harry is in Math. Ron's score was 1.3 standard deviations above the mean in English, whereas Harry's score was only 1 standard deviation above the mean in math.

Module 9

Learning Objectives

- Know the effect of score transformations on the percentage of cases falling above, below, or between various transformed scores

- Know the effect of score transformations on the mean and standard deviation of the set of scores

- Know the values of the mean and standard deviation for common standardized scores

- Convert scores from one type of standardized score to another

Module Summary

- You may need to adjust the scores in a distribution so that they are all higher, lower, more spread out, or more bunched up. This process is called a *transformation*. One type of transformation is to add a constant to every score in the distribution. Doing this would increase the mean by that constant as well. Alternatively, you may multiply every score by a constant. This would cause the mean to be multiplied by that constant as well.

- Transformations also affect measures of dispersion. Adding a constant to every score in the distribution would not affect the standard deviation. This is because this transformation doesn't move the scores closer to or further from the mean. Rather, this process shifts the entire data set in a specified direction. However, if we were to transform our data by multiplying every score by a constant, we could expect the standard deviation to be multiplied by that constant as well. In multiplying the scores, we are moving each score further away from the mean, so we can expect the standard deviation to increase.

- Regardless of the method of transformation that you use on your data set, scores that are a certain number of standard deviations from the mean will always fall in the same place on a normal curve.

- Standardized scores are actually a transformation. You change any standardized score back to its raw score with the following formula:

$$Raw = s * z + M$$

Computational Exercises

1. A psychology teacher realizes that a recent test was very difficult and decides to give everyone in the class an additional 8 points. If Sana initially received a grade of 78 and the class had an $M = 72$ with an $SD = 4$, what is her standardized score? How does this transformation impact the descriptive statistics for this class and for Sana?

Here are the numbers of court cases a particular judge saw per day for the past week. Use this information for questions 2-5.

$$7, 5, 9, 3, 4$$

2. What are the mean and standard deviation for these cases?

3. The following week the judge is inundated with work, as his case load is doubled! What are the mean measures of central tendency and dispersion for the judge during this week?

4. The following week, the judge continues to have a busy time at work. However, it isn't as bad as the previous week (the week in which he had double the cases; question 3) because he has 2 fewer cases per day. What are the measures of central tendency and dispersion for this week? (Using the original scores, multiple each score by 2 and then subtract 2 from each score.)

5. Using your information from question 3, compare the busiest day (the day with the most cases) to the busiest day he had during the original week. Which day was more different from the average? How much more in terms of percentage of the normal curve?

6. If we were to obtain another sample of 50 participants and each of them provided the exact same responses as the ones we had obtained from the first sample (so all frequencies are now doubled), what would the measures of central tendency become? Which was most affected by this change?

Computational Answers

1. This will increase the mean of the scores and Sana's score, but it will not affect the standard deviation. Sana's grade will increase, but her standardized score will remain the same at 1.50.

2. The mean would be 5.6, and the standard deviation would be 2.15 (using N) or 2.41 (using $n - 1$).

3. The mean for this week would be 11.20, and the standard deviation would be 4.30 (using N) or 4.82 (using $n - 1$). Each of these scores is multiplied by two.

4. The mean would decrease by 2 to 9.20. The standard deviation would be unaffected and remain at 4.30 (using N) or 4.82 (using $n - 1$).

5. The z score for the busiest day with the original scores is 1.58. The z score for the busiest day from the next week is also 1.58. Even though the scores changed, their positions in the distribution remained the same.

6. None of the measures of central tendency would be affected by doubling the number of scores.

True/False Questions

1. Converting all of the scores in a distribution to z scores automatically converts the distribution to a normal distribution.

2. The average golf score for a golf team was 4. It was later found out that the team was cheating on their scores and all of the members were reducing their scores by 3 strokes. The true mean of the golf team was 7.

3. Multiplying all of the scores in a distribution by a constant will cause the mean to be multiplied by that constant but not affect the measure of dispersion.

4. The rules of transformation do not apply to distributions that are not normal.

5. Dividing all of the scores in a distribution by the same number will decrease the variability in the distribution.

True/False Answers

1. False

2. True

3. False

4. False

5. True

Short-Answer Questions

1. How are the mean and standard deviation affected by adding a constant to all the scores in a data set?

2. How are the mean and standard deviation affected by multiplying all the scores in a data set by a constant?

3. What does adding a constant to each score in a data set do to z scores? Why is this?

4. What does multiplying each score by a constant do to z scores? Why is this?

5. You design a measure to assess your classmates' opinion of this course, ranging from –10 to 10. However, the teacher asks to see the scores, and you don't want to hurt his feelings by

showing that some of the students provided negative ratings. How could you fix this problem while not actually changing the location of the scores in the normal distribution?

Answers

1. The mean will be changed by the same amount as the constant. The standard deviation is unaffected by adding a constant.

2. The mean and standard deviation are both multiplied by that constant.

3. Adding a constant to each score would not change any of the z scores. This is because adding a constant "shifts" the entire distribution in one direction. This shift does not affect the location of the scores in relation to the mean, so their differences in the numerator remain the same. The standard deviation in the denominator also remains the same.

4. The z scores do not change. Although the variability within the distribution will increase, the standard deviation also will increase. Therefore, the location of each score relative to the mean does not change.

5. You could add a constant of 10 to every score. This would change the scale to 0-20, which makes all of the scores positive but does not change their location in relation to the mean.

Multiple-Choice Questions

1. Company B needs to increase its profits. As a result, it increases the price on all of its models by $10. What is the new mean of its stereos?

Company	Mean	SD
A	450	23
B	375	12
C	475	57

   a. 375   c. 365

   b. 385   d. Need more information

2. A group of students are misbehaving, so their teacher states that they will all lose 5 points on their next test. If the students all earned an average grade of 72 with a standard deviation of 4, what would the new mean and standard deviation be with the penalty?

   a. $M = 72$; $SD = 9$ c. $M = 67$; $SD = 9$

   b. $M = 67$; $SD = 4$ d. $M = 72$; $SD = 4$

3. You are playing a number game with a friend. You give her a list of 10 numbers that has an $M = 62$ and $SD = 4$. You then tell her to add 4 to every number and then multiply every number by 2. She is amazed when you immediately tell her the new mean and standard deviation after making these changes. What are these new values?

   a. $M = 66$; $SD = 8$ c. $M = 132$; $SD = 8$

   b. $M = 102$; $SD = 12$ d. $M = 132$; $SD = 16$

4. Your friend thinks you cheated at the number game you played in the previous question. To prove her wrong, you tell her you are willing to do it again and will even let her provide you with 20 random numbers. The descriptive statistics of these new numbers are $M = 70$ and $SD = 15$. To make it extra hard, your friend tells you that you should first divide all the numbers by 5, then multiply them all by 3, and finally subtract 6 from all scores. You astonish your friend by telling her that the mean and standard deviation would be

   a. $M = 42$; $SD = 3$ c. $M = 90$; $SD = 7$

   b. $M = 36$; $SD = 9$ d. $M = 14$; $SD = 5$

5. You own a small dog who eats an average of 12 oz. of food per day with an $SD = 2$. One day while driving home, you encounter a similar dog that has been abandoned. Unable to leave this dog behind, you take it home. If this new dog has the same eating habits as the one you currently own, what can you expect the statistics to be for the food consumed by both dogs?

a. $M = 12$; $SD = 2$

c. $M = 24$; $SD = 2$

b. $M = 14$; $SD = 4$

d. $M = 24$; $SD = 4$

Multiple-Choice Answers

1. B

4. B

2. B

5. D

3. C

Module Quiz

1. You have read a recent report that drinking 2 caffeinated beverages will provide double the effects of drinking just 1. If the average person reports an increase of $M = 10$, $SD = 3$ in energy after drinking 1 beverage, what will be the statistics from drinking 2?

2. Unfortunately, drinking more than 2 caffeinated beverages has diminishing returns. Participants in a new study reported that drinking 3 beverages increased their average energy by only 4 more points beyond the increase for 2 caffeinated beverages, and drinking 4 beverages increased their average energy by only 2 more points beyond the increase for 3 caffeinated beverages. What would the mean and standard deviation of those who consumed 3 caffeinated beverages and 4 caffeinated beverages?

3. Why does multiplying scores in a distribution by a constant impact measures of variability but adding a constant to them does not?

4. Which of the following transformations has the greatest chance of turning scores at the edge of a distribution into an outlier: division, multiplication, addition, subtraction?

5. When teachers claim they are "curving" grades on a test by increasing everyone's grades by a set number of points, what in fact are they doing?

Quiz Answers

1. You can expect an $M = 20$ and $SD = 6$.

2. 3 beverages: $M = 24$, $SD = 6$; 4 beverages: $M = 26$, $SD = 6$.

3. Multiplying scores in a distribution by a constant increases the distance between the scores, whereas adding a constant does not. Increasing the distance between scores affects measures of variability.

4. Multiplication. This is the only transformation in which scores are pushed further from the mean.

5. They are adding a constant to all the grades to improve the mean. This is not "curving" that would involve fitting all the grades to a normal curve.

Module 10

Learning Objectives

- Distinguish between mutually exclusive, embedded, and overlapping outcomes on a single trial

- Distinguish between independent and dependent outcomes in a series of trials

- Know the conditions under which the addition theorem applies

- Know the conditions under which the multiplication theorem applies

- Distinguish between theoretical and empirical outcomes

- Understand the use of empirical versus theoretical data in making inferences

Module Summary

- The properties of the normal curve and skewed curve allow us to perform most inferential statistics. Probability theory helps us to understand the properties of the normal curve and skewed curve. Therefore, it is important to have an understanding of probability as it underlies these curves. However, probability is a vast topic, and this book will touch on only a few key elements that are relevant to our statistics.

- *Probability* is the chance that you will obtain a specific outcome (e.g., outcome A) out of many different outcomes (e.g., outcomes A, B, C, or D). Probability is frequently expressed as a proportion, representing the chances of obtaining the specified outcome divided by the total number of outcomes. In the previous example, the probability of obtaining outcome A, expressed as $p(A)$, would be 1/4. This is because there is one outcome A, but four possible outcomes.

- The previous example assumes that the chance of getting any particular outcome (A, B, C, or D) is equal. This is referred to as an *equally likely model*. Not all situations conform to the

94

equally likely model. For example, if there were two outcome A's (A, A, B, C, D), then the chances of getting an A are higher than the chances of getting another letter. However, the majority of statistics for this text are based on the equally likely model.

- *Mutually exclusive outcomes* indicate that you can obtain only one outcome per trial. If the possible outcomes are A, B, C, D and you obtain outcome B, then you cannot have an outcome of A, C, or D on that trial. Although it is possible for outcomes to not be mutually exclusive, the majority of the statistics in this text are based on mutually exclusive outcomes.

- The *addition theorem* states that the probability of either of two outcomes on a *single trial* is equal to the sum of the two individual probabilities. Returning to our example of A, B, C, D, the probability of obtaining either an A or B on a single trial is equal to the probability of obtaining A plus the probability of obtaining B. Therefore, this probability would be 1/4 + 1/4 = 1/2. Look for the word "either" when using the addition theorem.

- Another consideration in probability is how the outcome of one trial impacts the outcome of another trial. *Independent outcomes* are those in which the outcome of one trial has no impact on the outcome of another trial. If you were to flip a coin twice, you would have a 50% chance of obtaining a heads on the second flip, regardless of what side the coin landed on in the first slip. The independence of outcomes on successive trials will play a role in which type of statistical test you will use. This will be covered at length in later chapters.

- Finally, the *multiplication theorem* states that the probability of obtaining two specified outcomes in two *different trials* is the product of their individual probabilities. Using the A, B, C, D example, this means that if the probability of obtaining A on the first trial is 1/4 and the probability of obtaining B on the second trial is 1/4, then the probability of obtaining A and then B would be 1/4 * 1/4 = 1/16. Look for the word "and" when using the multiplication theorem.

- When you have only a limited number of trials, you cannot guarantee any certain outcome, regardless of its probabilities. In other words, even though we know what to expect, that might or might not actually happen. Probabilities are *theoretical*, meaning probabilities are what you expect to happen. Actual results are *empirical*, meaning they actually happen. However, as you increase the number of trials, your empirical findings will more closely resemble your theoretical expectations. If we performed a trial with four equally likely outcome (A, B, C, D) 200 times, you could expect close to 50 A's, 50 B's, 50 C's, and 50 D's (although you would not expect exactly these numbers).

- Research in the social sciences frequently focuses on comparing two different groups, such as a treatment and a control (no-treatment) group. Using the equally likely model, you would expect that there to be no difference between the groups, that the scores of the treatment group = scores of the control group. However, if the treatment is more effective, then you would notice a difference between these groups. This difference would mean that the probability of falling in one specific group is not equally likely; it depends on the person's treatment. Such as unequal outcome suggests that the treatment is effective.

Computational Exercises

1. Samantha wants to rent a comedy from the movie store. In browsing the comedy section, she sees four movies that she would really like to see but is uncertain of which one to pick. She calls her roommate and asks for her opinion on the matter. If a third friend, unknown to Samantha and her roommate, had picked one of the four titles, and then Samantha were to pick a title but not tell her roommate (so that the selections they both would make would be

mutually exclusive), what are the chances that both Samantha and her roommate would pick the same title that the third friend picked?

2. You are attending a conference in which there is a representative from each one of the 50 United States (you are not a representative). You are introduced to one of the representatives. What is the probability that: A. This person is from New York? B. This person is from California, Texas, or Rhode Island?

3. You are playing a dice game with your friend that uses 3 six-sided die. A. What is the probability that you will roll a 1 on with the first die, a 2 with the second, and a 3 with the third? B. What is the probability that you will roll a 6 with the first die, a 6 or a 4 with the second die, and a 1 with the third die? C. What is the probability that will roll a value greater than 3 on the first die, a value of 3 less on the second, and a 4 on the third?

4. You and a few of your friends are preparing for the beach and discussing which color beach towel each person should bring. You have a total of 15 beach towels: 3 are blue, 2 are green, 6 are red, and 4 are pink. A. If you close your eyes and grab a towel, what is the probability that it will be red or pink? B. What is the probability that you took a green towel and that your friend (who also takes a towel without looking) will take the other green towel?

5. Sam and Brenda are going to adopt a cat. They go to a shelter and see 4 orange tabbies, 3 calicos, 2 Siamese, and 5 grey tabbies. A. What is the probability that they adopt a tabby cat? B. What is the probability they adopt a grey tabby or a Siamese cat?

Computational Answers

1. Chance of picking the same title: $\left(\dfrac{1}{4}\right)\left(\dfrac{1}{4}\right) = \dfrac{1}{16}$.

2. A. 1/50. B. 3/50.

3. A. $\left(\dfrac{1}{6}\right)\left(\dfrac{1}{6}\right)\left(\dfrac{1}{6}\right) = \dfrac{1}{216}$.

B. $\left(\dfrac{1}{6}\right)\left(\dfrac{2}{6}\right)\left(\dfrac{1}{6}\right) = \dfrac{2}{216}$.

C. $\left(\dfrac{3}{6}\right)\left(\dfrac{3}{6}\right)\left(\dfrac{1}{6}\right) = \dfrac{9}{216}$.

4. A. $\left(\dfrac{6}{15}\right)+\left(\dfrac{4}{15}\right) = \dfrac{10}{15}$.

B. $\left(\dfrac{2}{15}\right)\left(\dfrac{1}{14}\right) = \dfrac{2}{210}$.

5. A. $\left(\dfrac{4}{14}\right)+\left(\dfrac{5}{14}\right) = \dfrac{9}{14}$.

B. $\left(\dfrac{2}{14}\right)+\left(\dfrac{5}{14}\right) = \dfrac{7}{14}$.

True/False Questions

1. The probability of all possible outcomes must sum to 1.

2. Mutually exclusive outcomes indicate only one outcome is possible per trial.

3. Two soccer teams are playing each other. One of the teams has much better players than the other and is favored to win. The probability of this team winning is as equally likely as the team with the weaker players.

4. The probability that one outcome will occur or another outcome will occur is the product of these two individual probabilities.

5. If the probability that you will pass a math test is 0.90 and the probability you will pass a history test is .80, the probability that you will pass both tests is 0.72.

6. The addition theorem applies to outcomes over many trials.

7. You and a friend are drawing straws from a bag that contains 6 straws. You draw one straw and keep it. Then, your friend draws one straw and keeps it. Then you draw again and so on until there are no straws left. These trials each have independent outcomes.

8. Probabilities are empirical, whereas observed results are theoretical.

9. Theoretical principles are used as standards by which empirical results are compared.

10. Statistics and probability will provide you with a definite understanding of why an outcome occurred.

True/False Answers

1. True	5. True	9. True
2. True	6. False	10. False
3. False	7. False	
4. False	8. False	

Short-Answer Questions

1. What does a probability of 0.75 indicate?

2. What are mutually exclusive outcomes? Are mutually exclusive outcomes common or rare when using statistics for the social sciences?

3. Compare and contrast when the addition theorem and the multiplication theorem should be used.

4. Sam is learning to play a song on the saxophone. Each day he practices for 2 hours. Is the probability that he will correctly the play song each day an independent outcome? Why or why not?

5. If you were conducting an experiment in which you wanted to obtain the public's opinion on whether they enjoyed a recent movie, why would it be important to have independent outcomes? (Hint: consider each person you would ask as a separate trial.)

6. What does it mean for something to be empirical as opposed to theoretical?

7. What does the expression "9 times out of 10 . . ." really indicate?

Answers

1. It indicates that theoretically, one set of outcomes occurs 75% of the time and another 25% of the time.

2. Mutually exclusive outcomes are those in which only one outcome is possible per trial. These are common when using statistics from the social science perspective.

3. The addition theorem should be used for a single trial, when assessing mutually exclusive outcomes. The multiplication theorem is used when assessing multiple trials with independent outcomes.

4. These trials are not independent. Each day that Sam practices, his ability to play the song improves. Thus, the probability that he will correctly play the song today depends upon how well he played the song the previous day (assuming he is improving in his playing).

5. Having independent outcomes would ensure that each opinion is not biased by another person's opinion. In other words, you want to ensure that each person's response is not the result of a previous person's response. This would bias your overall poll.

6. Empirical refers to observed results, whereas theoretical refers to expected results based on an ideal situation.

7. This would suggest that the probability of obtaining a specific result is 0.9. In other words, it indicates that the chance of obtaining one specific outcome is rather high.

Multiple-Choice Questions

1. If the probability of an event occurring is .45, what is the probability of the event not occurring?

      a. 0.45   b. 0.35

c. 0.50   d. 0.55

2. If the probability of an event occurring is .1 and the probabilities of alterative outcomes are

    equally likely, how many possible outcomes are there?

    a. 1   c. 5

    b. 2   d. 10

3. What is the probability that you will select the correct choice for this question without

    knowing what you are being asked?

    a. 0.5   c. 0.25

    b. 0.1   d. 1

4. There are 20 questions in a section of a test. If you were to select a choice completely at

    random for all 20 questions, what is the probability that you would get all of the answers

    correct?

    a. 0.25   c. 0.01

    b. 0.1   d. < 0.001

5. If you draw a card from a standard deck of 52 cards, which of these outcomes is not mutually

    exclusive?

    a. It is a high card or a low card.   c. It is a king or a queen

    b. It is a diamond or a heart.   d. It is a king or a heart.

6. If you were to draw a card from a deck of 52 cards, what is the probability that it would be a

    face card (jack, queen, king)?

    a. 0.23   c. 0.06

    b. 0.02   d. 0.01

7. The multiplication theorem cannot be applied under which circumstances?

a. Single trials

c. Independent outcomes

b. Mutually exclusive outcomes

d. Dichotomous outcomes

8. On a multiple-choice question, there are 5 options. If you are able to eliminate 2 of the possible choices, what is the probability of selecting the correct answer from the remaining choices?

a. 0.20

c. 0.50

b. 0.33

d. 0.75

Multiple-Choice Answers

1. D

5. D

2. D

6. A

3. C

7. A

4. D

8. B

Module Quiz

1. You recently ordered 4 replacement parts for a lamp you purchased. The sales representative tells you that the parts will either come next Wednesday or next Thursday. What is the probability that parts come on Tuesday or Wednesday? Wednesday or Thursday?

2. Elba was asked by her roommate to pick up some tomato sauce at the grocery store. Unfortunately, she forgot what brand she was supposed to purchase. The grocery store has 10 different brands of tomato sauce. What is the probability that Elba will pick the correct brand? If Elba were to buy 2 brands, what is the probability that at least one of them will be the correct brand?

3. What does it mean for outcomes to be equally likely? Can there be more than 2 equally likely outcomes?

4. Anita is a clumsy waitress and has a tendency to drop things. If the probability that a cup will fall on its side is .5, right side up is .25, and upside down is .25, what is the probability that the 3 cups Anita just dropped will all land right side up?

5. Would you consider your performance on tests in this class as independent outcomes? What about your performance on tests in this class as compared to tests in other classes?

Quiz Answers

1. Tuesday or Wednesday: $\frac{0}{2} + \frac{1}{2} = \frac{1}{2}$; Wednesday or Thursday: $\frac{1}{2} + \frac{1}{2} = \frac{2}{2}$.

2. The probability she will pick the correct brand is 1/10 or .1. The probability she will pick one of the correct brands if she bought 2 is 2/10 or .2.

3. It means that the probability of obtaining one outcome is as likely as obtaining the other outcome. There can be multiple equally likely outcomes. For example, the probabilities for four equally likely outcomes would be .25.

4. 0.0156.

5. Your performance on tests in this class should not be considered as independent outcomes, as your ability to do well in statistics will improve with each successive trial. Your performance across classes could be considered independent outcomes, as your knowledge of statistics may not influence your knowledge of history. However, an argument for them being not independent could be made, as study skills you develop for statistics may carry over to other classes.

Module 11

Learning Objectives

- Calculate probability by listing all possible outcomes

- Calculate probability by expanding the binomial formula

- Find probability by using a binomial table

- Understand the relationship between likelihood in a dichotomous model and the shape of the outcome distribution

Module Summary

- **Dichotomous outcomes** are situations that have only two possible outcomes. The results from dichotomous outcomes with a probability of 0.5 per outcome tend to be distributed in a pattern that resembles the normal curve. Recall that p represents the probability an outcome will occur and q represents the probability an outcome will not occur.

- This pattern is referred to as the binomial distribution. By using this pattern, it is possible to determine the probability that you would obtain any possible combination of outcomes across any amount of trials. There are two methods to determine these probabilities. The first requires you to add all of the probabilities of obtaining a combination with your desired number of outcomes. For example, if you wanted to determine the probability of obtaining at least two heads in three coin flips, you would add the probability of all ways to earn exactly two heads and then all the ways of earning three heads. This sum would provide you with the probability of earning at least two heads in three flips.

- However, the previously mentioned process is rather tedious. An alternative method is using the binomial formula, which is $(p + q)^N$. When using this formula to calculate probabilities, it is necessary to expand the formula through the use of a process called ***binomial expansion***. This process expands the formula to appear as follows:

$$(p+q)^N = p^N + Np^{N-1}q + \frac{N(N-1)}{1 \times 2} p^{N-2}q^2 + ...q^N$$

This formula provides the probability of obtaining all combinations of outcomes in N trials.

- If you were to graph the results of this formula, it would bear a strong resemblance to the normal curve. The difference is that a normal curve is for continuous data, and a binomial table is for discrete data. With large sample sizes, we can use the normal curve instead of the binomial table even though the data are discrete.

- Aside from binomial expansion, the text provides a table (Appendix B) that will provide you with the probabilities for obtaining each specific outcome for a number of trials (up to 10). Also, the table provides information for a number of different probabilities for p and q.

- The binomial distribution is relevant to the social sciences, and specifically experimentation, because it provides a set of expected probabilities from an unlimited amount of trials. This means that you do not need to conduct an experiment hundreds of times in order to feel confident about the results. Rather, you can consult the normal curve to determine the probability of obtaining such a result. For example, you are interested in determining if a group of people are generous. You decide to ask 10 people to donate to a charity, with the expected response of either a yes or no. Also, you expect that the probability of a person responding with a yes to be 0.5. It would be highly unlikely that you would obtain 10 yes responses. However, if you did, you may conclude that the chance of obtaining a yes response from this group is higher than expected. You could further conclude that the reason you got 10 yes responses is because the probability of obtaining a yes response is much higher than the 0.5 that you had originally thought.

Computational Exercises

1. At the beach, you encounter a person playing a game. He has two cups with an almond under one of them. He shuffles the cups and asks you to pick which one has the almond. He does this three times (three trials). However, he shuffles the cups so quickly that you lose track of the almond pretty quickly and now have to guess at random on each of the four trials. What is the probability for all of the combinations? What are the p and q for each combination?

2. Sketch a graph of the probabilities for getting a hit across the three trials in question 1. If the person with the cups offers you a prize for guessing it three times in a row but states that you must pay him if you miss all three times in a row, is this a fair deal?

3. You recently ordered 6 replacement parts for a lamp you purchased. Each part will be shipped independently of the other parts. The sales representative tells you that each part will come either next Wednesday or next Thursday, and all parts have an equal chance of coming on either day. For your needs, Wednesday is a hit, Thursday is a miss. Using binomial expansion, what are the probabilities of getting exactly 4 of the 6 parts on Wednesday rather than on Thursday?

4. Surprisingly, you do not receive any of your parts on either day (from the previous question). Upset, you call the company and ask when you should expect them. The company says that you should expect all 6 parts either Monday or Tuesday. However, the company has mailed with an express delivery and there is a greater chance (0.75) you will obtain them on Monday. What is the probability of obtaining all 6 parts on Monday using the table in Appendix B?

5. Complete a binomial expansion for 7 trials, where each trial has two possible outcomes with equal 0.5 probabilities.

6. David is practicing his dart throwing on a dart board whose area is 10% bull's-eye and 90% miss. If he isn't very good and is just doing his best to hit the dart board (which he always will), what is the probability he will hit the bull's-eye at least twice in 7 throws? (Use the table in appendix B.)

7. Using the example from above, David has improved his accuracy and now has a 20% chance of hitting the bull's-eye. If he throws 5 times, what is the probability that he will hit the bull's-eye once, twice, or three times?

8. David feels so confident in his dart throwing ability (now 30% of hitting the bull's-eye), he decides to impress his friend Shakra. He tells Shakra that she must pay him a dollar if he can hit the bull's-eye at least twice in eight throws; if he doesn't, he will pay her a dollar. Should Shakra take this bet? Use the table in Appendix B to help you make the decision.

Computational Answers

1. Using "H" for "Hit" and "M" for "Miss":

Combination	p	q
MMM	0.125	0.875
HMM	0.125	0.875
MHM	0.125	0.875
HHM	0.125	0.875
MMH	0.125	0.875
HHH	0.125	0.875
MHH	0.125	0.875
HMH	0.125	0.875

2.

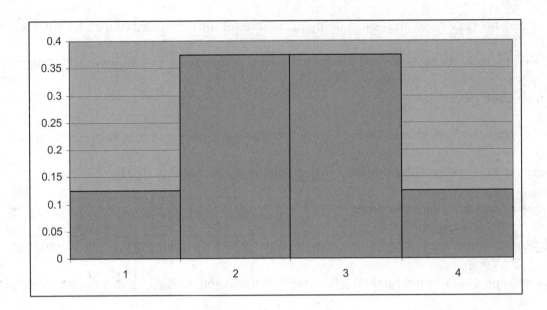

This is a fair deal because there is an equal chance of both outcomes. However, both situations are rare.

3.

$$0.5^6 + 6(0.5^5)(0.5) + \frac{(6)(5)}{(1)(2)}(0.5^4)(0.5^2) + \frac{(6)(5)(4)}{(1)(2)(3)}(0.5^3)(0.5^3) + \frac{(6)(5)(4)(3)}{(1)(2)(3)(4)}(0.5^2)(0.5^4) +$$

$$\frac{(6)(5)(4)(3)(2)}{(1)(2)(3)(4)(5)}(0.5^1)(0.5^5) + \frac{(6)(5)(4)(3)(2)(1)}{(1)(2)(3)(4)(5)(6)}(0.5^6) = 0.016.$$

Exactly four is the term: $\frac{(6)(5)}{(1)(2)}(0.5)^4(0.5)^2 = 15(0.0625)(0.25) = 0.234$.

4. $N = 6$, $p = 0.75$, $q = 0.25$, $p(6) = 0.178$.

5.

$$0.5^7 + 7(0.5^6)(0.5) + \frac{(7)(6)}{(1)(2)}(0.5^5)(0.5^2) + \frac{(7)(6)(5)}{(1)(2)(3)}(0.5^4)(0.5^3) + \frac{(7)(6)(5)(4)}{(1)(2)(3)(4)}(0.5^3)(0.5^4) +$$

$$\frac{(7)(6)(5)(4)(3)}{(1)(2)(3)(4)(5)}(0.5^2)(0.5^5) + \frac{(7)(6)(5)(4)(3)(2)}{(1)(2)(3)(4)(5)(6)}(0.5^1)(0.5^{06}) + \frac{(7)(6)(5)(4)(3)(2)(1)}{(1)(2)(3)(4)(5)(6)(7)}(0.5^7).$$

6. $N = 7$, $p = 0.10$, $q = 0.9$, $p(\geq 2) = 0.150$.

7. $N = 5$, $p = 0.20$, $q = 0.8$, $p(1) = 0.4096$; $p(2) = 0.2048$; $p(3) = 0.0512$. The total probability is 0.666.

8. $N = 8$, $p = 0.30$, $q = 0.7$, $p(\geq 2) = 0.745$. She should not take the bet.

True/False Questions

1. Dichotomous outcomes always have at least three outcomes.

2. If the probability of two outcomes is equally likely, then binomial distribution resembles the normal distribution after many trials.

3. If the probability of two outcomes is not equally likely, then the binomial distribution will appear skewed.

4. You and a friend are playing catch. If your friend has a 0.4 chance of missing a catch, then the probability of him missing the next two throws is 0.8.

5. The majority of events in life are dichotomous with equally likely outcomes.

True/False Answers
1. False   3. True   5. False

2. True   4. False

Short-Answer Questions

1. What are three criteria must be met in order for a set of outcomes to be considered dichotomous?

2. Using the binominal distribution, how does the probability change as you increase the number of "hits" from 0 if the outcomes are equally likely?

3. Why is the binominal distribution important to other inferential statistics?

4. When would a set of dichotomous outcomes produce a skewed binomial distribution?

5. Are the following dichotomous situations equally likely or not? A) Winning a game at a carnival or losing the game. B) Passing a test or failing a test. C) Enjoying a movie or not enjoying it. D) Flipping a coin and getting a head or getting a tail. E) Dropping a piece of toast butter side up or dry side up.

<u>Answers</u>

1. The three criteria are there must be 1) two possible outcomes that are 2) mutually exclusive and 3) independent.

2. The probability increases as you increase the number of hits from 0. However, after obtaining approximately an equal number of hits and misses, the probability of obtaining more hits than misses decreases. Eventually, the probability of obtaining all hits is equal to the probability of obtaining no hits.

3. It is important because when the outcomes are equally likely, the distribution appears normal. This enables us to determine whether or not an observed set of outcomes, such as seeing a large difference between groups, is highly probable (meaning that you would expect this difference) or highly improbable (meaning that you would not normally expect this difference). Based on this distinction, you may be inclined to determine whether or not a treatment "worked."

4. A skewed distribution is produced when the probabilities for a dichotomous outcome are not equal.

5. A) Not equally likely; B) Not equally likely; C) Not equally likely; D) Equally likely; E) Equally likely.

Multiple-Choice Questions

1. Which of the following is false: Dichotomous outcomes always

    a. Have two solutions   c. Are mutually exclusive

    b. Are equally likely   d. Are independent

2. The weatherman forecasts that the chance it will rain during the next 7-day week is 0.4 per day. Using the binomial table in appendix B, what is the probability it will rain for exactly 3 days?

a. 0.2903   c. 0.2160

b. 0.7009   d. 0.6000

3. Using the probabilities in question 2, what is the probability that it will rain for at least 4 days?

     a. 0.1935   c. 0.2897

     b. 0.2881   d. 0.2679

4. Using the probabilities in question 2, what is the probability it will rain for either one or three days?

     a. 0.2903   c. 0.2548

     b. 0.1306   d. 0.4209

5. A salesman states that his cough medicine is guaranteed to work! He reports that he gave the medicine to 7 people, and they all immediately lost their cough. You are skeptical of this salesman and want to tell him the probability that this would happen. What is the probability of this happening assuming that the medicine had an equal chance of curing the cough or not curing the cough?

     a. 0.0547   c. 0.1641

     b. 0.0078   d. 0.0098

6. The salesman in question 5 confesses that it didn't cure all of the coughs. However, he states that there is a 95% chance it would cure your cough. If you were to take this medicine 5 times and this probability is true, what is the probability that the medicine would cure your cough every time?

     a. 0.9514   c. 0.5054

     b. 1   d. 0.7738

7. As it turns out, the salesman from question 7 was lying about his cough medicine. In fact, it only helped relieve coughing 0.10 percent of the time. If you were to give this to 3 people, what is the probability that it would reduce at least one person's cough?

    a. 0.7290   c. 0.2430

    b. 0.2710   d. 0.0280

8. You are starring in a play that depicts a 1950's gangster who habitually flips a coin. You begin to count the number of times you obtain a tails in each flip. If your recent practice session included 4 flips, how many different ways would there be to obtain 2 or more tails?

    a. 9   c. 11

    b. 10   d. 12

9. Using the above example (Question 8), what is the probability that you would obtain 3 or more tails in 4 flips?

    a. 0.2500

    b. 0.3125

    c. 0.0675

    d. 0.3750

Multiple-Choice Answers

1. B   6. D

2. A   7. B

3. C   8. C

4. D   9. B

5. B

Module Quiz

1. At the beach, you encounter a person playing a game. He has two cups with an almond under one of them. He shuffles the cups and asks you to pick which one has the almond. He does this tree times (three trials). However, he shuffles the cups so quickly; you lose track the almond pretty quickly and now have to guess at random on each of the three trials. List the all of the possible combinations of selecting the almond across three trials.

2. You are comparing the effect of a new therapy for autism. At the start of the study, you state that if the treatment doesn't work, then those who receive the therapy will be no better than those who did receive the therapy (the theoretical difference in symptoms between the two groups based on probability would be 0). However, you discover that those who had the treatment have fewer symptoms than those who did not. Furthermore, the difference in symptoms between the groups is quite large and the probability of observing this difference is 0.03. Can you conclude that your treatment works? Why or why not?

3. How would a binominal distribution in which the probability of a hit is 0.9 appear after an infinite number of trials?

4. Dichotomous outcomes are usually going to be on which scale of measurement?

5. Why is the binomial distribution so useful in statistics? (HINT: What happens to the binomial distribution as the number of trials increases?)

Quiz Answers

1. Almond = H, No Almond = M.

    MMM; HMM; MHM; MMH; HHM; HHH; MHH; HMH.

2. You can feel confident that your treatment had an effect on symptoms, but you cannot be certain that it worked. The fact that the chance of obtaining such a difference is very low does indicate provide strong evidence that the treatment worked; however, it does not prove

that it was effective because you have not tested all of the people with autism with your treatment.

3. Negatively skewed.

4. Nominal.

5. The binomial distribution begins to represent the normal curve, thus allowing us to use probabilities associated with the normal curve. Therefore, we do not need to carry out an unlimited number of trials to determine what the odds are for a certain number of outcomes. We can just use the probabilities associated with the normal curve.

Module 12

Learning Objectives

- Distinguish between various sampling methods
- Distinguish between types of variables
- Distinguish between null and alternative hypotheses
- Distinguish between directional and nondirectional hypotheses
- Write various types of hypotheses
- Understand the relationship between a research question and the directionality of the hypothesis
- Understand the impact of extraneous or confounding variables on interpretation of a study's result

Module Summary

- *Descriptive* statistics are used to describe the participants in a sample. Descriptive statistics include means and standard deviations. *Inferential* statistics are used to learn (or infer) about a larger *population* through the use of a sample.

- In order to use inferential statistics, your sample must be *representative*, meaning that it has properties similar to that of the population. A representative sample should be similar in almost all respects to the population of interest.

- Representative samples are drawn through different *sampling* techniques. One of the most commonly used sampling techniques is simple random sampling. **Simple random sampling** is the process in which every person in the population has the opportunity to be included in the sample and those that are included are done so at random. **Systematic sampling** involves selecting every n^{th} person. For example, you might select every 10th person from a list of all possible participants. Another commonly used sampling method is stratified random sampling. **Stratified random sampling** involves dividing the population into categories and

115

then drawing random samples from those categories. In this process, the proportion of each category in the population is maintained in the sample. For example, if you knew that the population of your school was 60% male and 40% female, you would want your sample to contain 60% males and 40% females. *Cluster sampling* involves selecting among predetermined groups, such as classes or teams. Finally, *convenience sampling* is taking a sample from those subjects that are available for the study at that particular time or who volunteer to participate. Convenience samples may not provide you with a representative sample.

- In research, there are typically two main variables of interest. The *independent variable* is the variable that is manipulated by the researcher and expected to bring about a difference in another variable. The *dependent variable* is the variable that is believed to be changed by the independent variable. The DV is what is observed in a research study.

- Although the IV and DV are the primary variables of interest, there are two additional variables that can pose problems to research designs. *Extraneous or confounding variables* are those that coexist with the IV and have an unwanted effect on the DV. This may cause the researcher to incorrectly conclude that the IV caused a change in the DV.

- A *hypothesis* is an educated guess about what is expected to happen in a research study. Generally, these statements are worded in terms of the effect the IV will have on the DV. *Research or alternative* hypotheses indicate that the IV will have an effect on the DV. These hypotheses can take two forms, direction or nondirectional. *Directional* hypotheses indicate that the IV is thought to impact the DV in a specific direction (e.g., make the DV increase or make the DV decrease). *Nondirectional* hypotheses do not indicate a specific direction in which the IV may impact the DV. Rather, they state that the IV will cause the DV to change, meaning that the DV may increase or decrease.

- The opposite of the alternative hypothesis is the ***null hypothesis***, which states that the IV will have no effect on the DV. In other words, the null hypothesis states that there will be no difference in the DV, regardless of what is done with the IV.

- The null and alternative hypotheses provide a dichotomous situation for all research questions: Either the IV does not have an effect on the DV (the null), or the IV does have an effect on the DV (the alternative). Research can never prove the alternative to be true because research will never be able to accurately assess an entire population. Thus, you can only say that you have not supported the null. In doing so, you have supported the alternative hypothesis.

Computational Exercises

1. A teacher is interested in obtaining feedback from his students regarding their opinion of the class. However, he has more than 500 students in the class, so he decides to take a sample. In order to avoid bias, he divides his class into sections based on their grades and samples 10 students from each section. What type of sampling strategy is the teacher using?

2. A researcher is interested in seeing how a new type of drug affects sleep in rats. The researcher gives some rats the drug and other rats no drug. The researcher then monitors their sleep. What is the IV and what is the DV in this study?

3. The researcher from the previous question finds that the rats with the drug tend to sleep more than those without the drug. She decides to see if this sleep effect can be offset by the presentation of light. Every rat is given the drug, and 1/2 are placed in a well-lit chamber,

while the other 1/2 are placed in a dark chamber. What is the IV and what is the DV in this study?

4. A developmental psychologist is interested in how children's math skills develop as they age. The psychologist monitors the development of math skills in a random sample of children as they progress from first to sixth grade. What is the IV and what is the DV in this study?

5. For the study proposed in question 3, what are the null and nondirectional alternative (research) hypotheses?

Computational Answers

1. This is a stratified sampling technique.

2. The IV is the drug, and the DV is the amount of sleep.

3. The IV is now the presence of light, and the DV is the amount of sleep.

4. The IV is grade or time or age, and the DV is math ability.

5. Null: The presence of light will not affect the amount of time spent sleeping by rats taking the drug. Alternative: The presence of light will affect the amount of time spent sleeping by rats taking the drug.

True/False Questions

1. Systematic random sampling involves randomly selecting participants from distinct categories within the population.

2. Joe decides to pick his softball team by selecting every other person available to play. This is a simple random sampling technique.

3. The goal of stratified sampling is to have the characteristics of your sample be similar to those of the population.

4. Convenience sampling is among the optimal methods of sampling.

5. Confounding variables are desirable in research studies.

1. False   3. True   5. False

2. False   4. False

Short-Answer Questions

1. Why are representative samples important when doing research? In general, what is the best method to use when trying to obtain a representative sample?

2. Why is convenience sampling not the optimal choice when choosing a sample strategy?

3. What is the purpose of the null hypothesis?

4. What is stated in the alternative hypothesis?

5. A scientist is interested in determining how a new type of chemical affects engine parts. The scientist places the chemical on 20 engines and compare how well these engines run in comparison to 20 engines without the chemical. Would the hypotheses for this study be directional or nondirectional? Why or why not?

Answers

1. Representative samples are necessary to draw valid conclusions about the population in which the samples originated. If the sample is not representative, the inferential statistical techniques that are used may be incorrect. The optimal method to obtain a representative sample is a simple random sampling.

2. Convenience sampling is using whatever subjects happen to be available. This method is not ideal because those that are available may not be representative of the larger population of study.

3. The null hypothesis serves a statistical/probability purpose. It represents what should be found if there was no effect of the IV. The purpose a research study is to find support for the alternative hypothesis.

4. That the independent variable will impact the dependent variable.

5. This study is nondirectional, as the scientist is uncertain of how the engine will be affected by the chemical.

Multiple-Choice Questions

1. Which type of sampling is most prone to giving you the least representative sample?

   a. Random sampling

   b. Stratified random sampling

   c. Convenience sampling

   d. Cluster sampling

2. Representative samples should

   a. contain participants that have similar characteristics to those in the population

   b. have sample statistics that are identical to population parameters

   c. be carefully handpicked from the population

   d. try to capture all of the variability in the population

3. In a study of language acquisition in primates, chimpanzees are randomly assigned to one of two groups. The first group receives language training with symbols, while the other group just interacts with humans. At the end of 1 month, the language acquisition of the chimpanzees is reviewed. However, the researchers notice that the chimps in the language group were slightly younger than the other group. What is the independent variable in this study?

   a. Language acquisition

   b. Age

   c. Time

   d. Language training or interaction

4. Using the study from question 3, what is a potential confounding variable?

   a. Language acquisition

   b. Age

   c. Time

   d. Language training or interaction

5. A team of scientists is interested in decreasing the time it takes for cookies to bake. They use a new food supplement in the cookie recipe and watch the baking time for the cookies. What is the null hypothesis in this study?

    a. The food supplement will change the time it takes for cookies to bake

    b. The food supplement will not change the time it takes for cookies to bake

    c. The food supplement will decrease the time it takes for cookies to bake

    d. The food supplement will not decrease the time it takes for cookies to bake

Multiple-Choice Answers

1. C	4. B
2. A	5. D
3. D	

Module Quiz

1. A photographer believes that black-and-white pictures are preferred over color pictures in a popular magazine. The magazine creates an opinion poll for their readers to provide the photographer with an answer. What are the IV and DV for this study? What are the null and alternative hypotheses?

2. You are interested in determining if a sports drink improves athletic performance. Is this a directional or nondirectional hypothesis?

3. Sampling error is not considered to be related to which of the following?

    a. The effect of the IV on the DV

    b. Random chance error

    c. Slight deviations between the sample statistics and population parameters

    d. The size of the sample

4. What is the best way to obtain a random sample?

5. The average test grade on a science test was 75 with a standard deviation of 6. If April received a grade of 82, what would be the alternative hypothesis if April wanted to determine if she scored significantly higher than the other students in her class?

6. Jessie is studying the effect of a new herbal supplement on memory. She gives some of her friends in her biology class the supplement. She then determines how well these friends remember information as compared to some of other friends in a chemistry class. What are the primary IV and DV of this study? What are potential confounding/extraneous variables in this study?

Quiz Answers

1. The IV is the type of photograph (color/black-and-white), and the DV is the preferences of the readers. Null: Black-and-white photographs are not preferred over color photographs. Alternative: Black-and-white photographs are preferred over color photographs.

2. It is a directional hypothesis because you are only interested in "improvements" in performance as opposed to declines in performance.

3. A.

4. Assign everyone in the population of interest a random number and then have a computer program, or some other random number generator, pick numbers.

5. Alternative: April did significantly better than other students.

6. The primary IV is the use of supplement, and the DV is the amount of information memorized. The potential confounding variable is class subject. The information in the biology class may be easier to memorize than that of the chemistry class (or vice versa). This could influence the results of the test.

Module 13

Learning Objectives

- Understand that probability, or being uncertain, always includes error

- Distinguish between two types of error—Type 1 and Type 2

Module Summary

- In a perfect situation, sample statistics would mirror population parameters. However, you can expect some difference between your sample statistics and your population parameters when conducting research. The extent to which these two measurements differ is referred to as *sampling error*. Sampling error is the deviation between a sample statistic and a population parameter that is attributed to chance, as opposed to other variables. Because this difference is attributed to chance, it is expected that the difference between a sample statistic and a population parameter will be small.

- In research, you manipulate your IV and expect to observe a change in the DV. Under the null hypothesis, we expect that there will be no change in the DV. In other words, the null hypothesis is supported when the deviation between a population parameter and a sample statistic is minimal, as this difference is thought to be attributed to sampling error. However, when we obtain a sample statistic that differs greatly from the population parameter, we reject the null hypothesis because we believe that the difference is caused by the manipulation of the IV. A large difference, one that is not expected by mere chance, is called a *significant difference*. However, this does not prove that the IV caused the difference. It is still possible that the large difference between the sample statistic and the population parameter is attributed to chance. This is the difficulty when working with probabilities.

123

- When you make a decision about whether to reject or retain (support) the null hypothesis, you may be making a correct decision (hit) or an incorrect decision (miss). As mentioned before, large differences between groups are thought to be related to the IV. If we obtain a large difference between our groups, and the difference was actually caused by the IV, then we have made a correct decision to reject the null (hit). However, we may also obtain a large difference due to chance. This would occur if our sample happened to be unrepresentative of the original (null) population. Unfortunately, we are unaware that the difference between our groups is due to chance, and so we would still reject the null. This situation is referred to as a *Type I (1) error*, where we have incorrectly rejected the null. Type 1 errors are also symbolized as *alpha* or *α*.

- Just as you may mistakenly reject the null in a Type 1 error, you may also mistakenly retain the null. It may be the case that you have selected a sample that is highly uncharacteristic of the treated (alternative) population. This sample may be more representative of the untreated (null) population, so the difference between the treated and untreated groups would be rather small. Remember, a small group difference is thought to be attributed to sampling error, so you would retain the null hypothesis. When the null hypothesis is retained incorrectly, you have committed a *Type II (2) error*. Type 2 errors are symbolized as *beta* or *β*.

- It is impossible to commit both a Type 1 and a Type 2 error simultaneously. However, there is always the chance of committing either a Type 1 or a Type 2 error, and this threat will be prevalent throughout this text and all hypothesis testing. Thus, the best way to feel confident about your results is to replicate them with many different samples.

True/False Questions

1. A significant difference refers to any difference between two samples.

2. Type 1 errors refer to instances when the null is incorrectly rejected.

3. It is possible to make both a Type 1 error and a Type 2 error with the same hypothesis test.

4. One possible cause of a Type 1 or 2 error is obtaining a sample that is unrepresentative of the population.

5. In retaining the null, you may have committed a Type 2 error.

True/False Answers

1. False   3. False   5. True

2. True   4. True

Short-Answer Questions

1. What are the potential causes of sampling error? How do these differ from the causes of a "real" (significant) difference between two groups?

2. What is a Type 1 error? What are the circumstances that would lead one to make this type of error?

3. What is a Type 2 error? What are the circumstances that would lead one to make this type of error?

4. Why would it be impossible to make both a Type 1 error and a Type 2 error simultaneously?

5. Are all differences between two groups meaningful differences? How could you go about determining whether or not a difference is in fact meaningful?

6. How does probability aid in making a decision of whether or not to retain or reject the null hypothesis?

Answers

1. Sampling error is potentially caused by random chance error that causes the sample statistics and population parameters to differ. This differs from "real" differences in that whereas sampling error is caused by chance, a real difference is caused by another variable (that the researcher hopes is the IV).

2. A Type 1 error is incorrectly rejecting the null hypothesis. This occurs when you obtain samples that are very different but their differences occurred by chance, not because of the effect of another variable. In reality, there was no effect of the independent variable.

3. A Type 2 error is incorrectly retaining the null hypothesis. This occurs when you obtain samples that are very similar, but their similarity is due to the samples being unrepresentative of their respective populations rather than an actual lack of difference between the samples. In reality, there was an effect of the IV.

4. Type 1 errors can be made only when the null is rejected, whereas Type 2 errors can be made only when the null is retained. You cannot both retain and reject the null.

5. No, not all differences are meaningful. Some are related to chance. You could determine whether or not the difference is meaningful by comparing the difference between your groups to the expected amount of difference (standardized error) if the null hypothesis were really true.

6. Probability indicates the odds of obtaining this observed value based on chance alone. If there is a large probability that we would obtain this value by chance, then we should retain the null. However, if there is small possibility that we would obtain this via chance, then we should reject the null.

Multiple-Choice Questions

1. A significant difference refers to

    a. A very large difference between two groups

    b. Any difference between two groups

    c. A difference between two groups that is greater than would be expected by chance

    d. A difference between two group means that is less than would be expected by chance

2. If you were to retain a true null hypothesis, then you would

    a. have committed a Type 1 error

    b. have committed a Type 2 error

    c. have made a correct decision

    d. the answer depends on the null and alternative hypotheses

3. If you were to reject a true null hypothesis, then you would

    a. have committed a Type 1 error

    b. have committed a Type 2 error

    c. have made a correct decision

    d. the answer depends on the null and alternative hypotheses

4. A Type 1 error can occur when

    a. Comparing two misrepresentative samples that were drawn from the same population and thus appear to be significantly different

    b. Comparing two representative samples that were drawn from the same population and thus appear to be similar

    c. Comparing two representative samples that were drawn from different populations and thus appear to be significantly different

d. Comparing two misrepresentative samples that were drawn from different populations and thus appear to be similar

5. You can be certain that you have committed a Type 2 error when

   a. You have a very large difference between your groups and you expected a very small difference

   b. You have a very small difference between your groups and you expected a very large difference

   c. Other studies have clearly indicated that the IV does impact the DV

   d. You can never be certain that you have committed a Type 2 error

Multiple-Choice Answers

1. C

2. C

3. A

4. A

5. D

Module Quiz

1. A photographer believes that black-and-white pictures are preferred over color pictures in a popular magazine because readers expressed a strong preference for black-and-white pictures during a poll. However, after publishing an issue with all black-and-white photographs, the magazine receives a number of angry letters from subscribers that state they love color photographs. These letters far outnumber the number of responses that they received in the original poll. Assuming that these letters fairly represent reader response, what type of error did the magazine publishers make in their conclusion from the original poll?

2. A research team is investigating the effects of a neuropeptide on alcohol addiction. They develop a study in which they are able to find a local group of individuals with this peptide

and a local group without the peptide. The researchers then determine the extent that these individuals are addicted to alcohol. They conclude that those with the peptide are in fact addicted to alcohol. It is later revealed that this peptide occurs only in those from this region of the country. Has an error been made, and if so, which type?

3. It has been consistently shown that exposure therapy is an effective treatment for specific phobias. You conduct a research study examining the effect of exposure treatment on participants with acrophobia (fear of heights) and find that the participants did not improve. You run to a colleague to tell them about your groundbreaking finding, but she disapproves. What is likely to be her argument against your finding?

4. Murray's parents are worried that he is sleeping too much because he sleeps about 9 hours per night. They base this on a comparison to his sister Peggy, who sleeps 6 hours a night. Unbeknown to the parents, the average person should sleep 8.5 hours a night, with a standard deviation of 1 hour. With this evidence, what type of error does it seem the parents made?

5. When can you be 100% certain when you reject or retain the null hypothesis?

Quiz Answers

1. The publishers committed a Type 2 error. Readers really do prefer color photographs, but the original poll did not find this preference.

2. There was no error committed because the researchers obtained a valid conclusion for their region with their sample. However, the findings may not generalize to other areas of the world. In other places, it would be assumed that they made a Type 1 error.

3. They will say that you probably committed a Type 2 error.

4. Murray's parents may have committed a Type 1 error, as Murray's sleep habits are within 1 standard deviation of the national average. Peggy is the unusual sleeper, as she is getting far less sleep than the national average.

5. You can never be 100% certain that you can reject or retain the null. There is always the chance for an error.

Module 14

Learning Objective

- Understand the logic underlying the numerator and denominator in most inferential test statistics

Module Summary

- The formula for z scores is the foundation for all types of inferential statistics. The numerator of this formula $(X - M)$ is comparing what you observed (X) to what you expected (M). The mean is the expected value because it is the most frequently occurring score when the data are normally distributed. The denominator (s) is the expected deviation of a score from the mean. In other words, the standard deviation is how much random error is expected between the observed score and the expected score.

- Using this interpretation of the z score formula, a z score could be expressed as a hypothesis test. The null hypothesis would state that you expect no difference (or minimal difference) between your observed score (X) and your expected score (M). You then determine how much of a difference you observed as compared to how much was expected (s). If this comparison tells you that you have a large difference, then you have grounds to reject the null hypothesis. If this comparison tells you that you have a small difference, then you would retain the null hypothesis. We can use probabilities to help us determine whether or not we have a big difference or a small difference. Using Appendix Table A, you can determine the probability that you would obtain this specific score. For example, an obtained z-score of 1.2 indicates that you would obtain a score higher than this 11.2% of the time and a score the same as or less than this 88.8% of the time. Based on these probabilities, you can decide whether or not the difference is large enough to reject the null hypothesis.

131

Computational Exercises

1. You are the teacher of a course. At the end of the term, one of your students who missed 5 days of class is upset that she had her final grade lowered as a result of her absences. She argues that she shouldn't be penalized for her absences because the other students had many more absences. In going over your class roll, you notice that the average student missed 2 days of class with a standard deviation of 1.7. What are the null and alternative hypotheses of this study? Is this study directional or nondirectional? How would you be inclined to respond to the student based on this information?

2. A z-score of 2.4 that is drawn from a population with a standard deviation of 14 corresponds to a raw score of 56. What is the expected value when drawing a raw score from this population? (Hint: Rearrange the z score formula to solve for M.)

3. A team of scientists is investigating the side effects of a new muscle enhancer on rats. They anecdotally discover that one of the rats given the muscle enhancer has become increasingly aggressive. This rat has bitten other rats 10 times in an hour, whereas the average rat bites only 3 times an hour, with a standard deviation of 1.5 bites. What are the null and alternative hypotheses? What is the probability that another rat would bite more than this rat? Would you reject or retain the null hypothesis?

Computational Answers

1. Null: The student was not absent more than other students. Alternative: The student was absent more than other students. This would be a directional study. Because the percentage of students missing more classes would be approximately 4%, I would state that this student's argument is wrong and that she should be penalized for missing class.

2. The expected value would be the mean, which is 22.4.

3. Null: The muscle enhancer does not increase aggressiveness in rats. Alternative: The muscle enhancer does increase aggressiveness in rats. The probability of another rat biting more than this rat is very close to 0. The researchers should be confident in rejecting the null hypothesis.

True/False Questions

1. The numerator of the z score formula makes a comparison between an obtained value and an expected value.

2. If you obtain a very high z score (> 4), then there is no chance that you have committed a Type 1 error.

3. The standardized random error provides a measure of comparison for the difference between the observed value and the expected value.

True/False Answers

1. True

2. False

3. True

Short-Answer Questions

1. How does the formula for a z score provide a template or prototype for all inferential statistics?

2. How does a z score of 0 support the null hypothesis?

3. Sarah received a grade of 1540 on the SAT. If the mean for the SAT that particular year was 1100 with a standard deviation of 100, would you believe that Sarah scored substantially better or worse than the population?

Answers

1. The z score formula provides a template in that you are determining the difference between an observed value and an expected value ($X - M$). Then you are comparing the difference between the observed and expected values to a standardized error, or the amount of difference you expected (s). If the difference between the values is similar to what was expected, then you can retain the null. Otherwise, the null is rejected.

2. A z score of 0 indicates that a sample mean is equal to the population mean. In other words, there is no difference between the sample and population, which is the null hypothesis.

3. Sarah scored 4.4 standard deviations above the average, substantially better than the population, as almost 100% of the population scored below her.

Multiple-Choice Questions

1. What does the denominator in the z score formula represent?

   a. An comparison between an observed value and an expected value

   b. An expected amount of random error expected from observed values

   c. The probability of obtaining this score by chance

   d. The standardized score of your obtained score

2. The average test grade on a science test was 75, with a standard deviation of 6. If April received a grade of 82, what percentage of the class did she do better than?

   a. 0.80   c. 0.88

   b. 0.82   d. 0.95

3. Using the above example, what would be the alternative hypothesis if April wanted to determine if she scored significantly higher than the other students in her class?

   a. April did significantly better than other students

   b. April did significantly different from other students

134

   c. April did not do significantly better than other students

   d. April did not do significantly different from other students

Multiple-Choice Answers

1. B

2. C

3. A

Module Quiz

1. The national average for an anxiety inventory is 30, with a standard deviation of 7. If you are a therapist and have been seeing a client who now reports a score of 24, what percentage of the population would score as low as or lower than your client?

2. Based on your response to question 1, would you believe that the client no longer has severe anxiety symptoms?

3. After speaking with your client about their surprising results, the client admits that he lied on the measure and is actually doing much worse. What type of error did you make in this situation?

Quiz Answers

1. 20%.

2. Yes, because the score is below the national average of people without an anxiety disorder.

3. A Type 1 error.

Module 15

Learning Objectives

- Understand the difference between a raw score distribution and a sampling distribution of the mean

- Understand why any sampling distribution of the mean is normally distributed

- Understand the impact of sample size on the shape of a sampling distribution

- Understand the impact of sample size on the size of the standard error of the mean

Module Summary

- If you graph the means of an infinite number of samples of any size from any distribution, the graph will appear to be normally distributed. This is the case regardless of the shape of the original distribution. This principle is referred to as the ***central limit theorem***. Also, as the size of each sample increases, the effect of central limit theorem becomes more pronounced. In fact, the means of samples with an $n > 30$ tend to rarely deviate from the center of the distribution.

- The distribution that is formed by the sample means is referred to as the ***sampling distribution of the mean***. The mean of the sampling distribution of the mean will always be equal to the population mean.

- The sampling distribution of the mean also contains a measure of the average deviation of any given sample mean from the population mean. Because we are now dealing with samples statistics instead of raw scores, the term that refers to the average deviation is the standard error rather than the standard deviation. Thus, the term for the average deviation of a sample mean from the population mean is referred to as the *standard error of the mean*.

The standard error of the mean is calculated by using the following formula:

$$\sigma_M = \frac{\sigma}{\sqrt{n}}$$

- The standard error of the mean will always be smaller than or equal to the standard deviation of the population. Furthermore, σ_M will become smaller as sample size increases. This is because as you increase the size of your sample, you are better able to approximate the population. In better approximating the population, you reduce the amount of sampling error you expect between your sample and the population. Conversely, σ_M will become larger as the sample size decreases until you have a sample of 1, in which case σ_M will be equal to σ.

Computational Exercises

1. Samples of $n = 16$ are drawn from a population with $\mu = 50$ and $\sigma = 10$. What is the standard error of the sampling distribution of the mean?

2. Using the information from question 1, change the sample size to $n = 24$. Now what is the standard error of the sampling distribution of the mean? Did the standard error of the mean increase or decrease from the previous example? Why?

3. The mean of a population is 25, and the standard deviation is 4. If you drew an infinite number of samples of $n = 12$, what sample mean would be 2 standard errors above the mean?

4. A population has a $\mu = 32$ and $\sigma = 10$. If you were to draw an infinite number of samples with an $n = 23$, what sample mean would be 1.5 standard errors above the mean?

5. A population has a $\mu = 54$ and $\sigma = 13.4$. If you were to draw an infinite number of samples with an $n = 8$, what sample mean would be 3.2 standard errors below the mean?

Computational Answers

1. $\sigma_M = \dfrac{10}{\sqrt{16}} = 2.5.$

2. $\sigma_M = \dfrac{10}{\sqrt{24}} = 2.04$. The standard error decreased because the sample size increased. This

    enables the sample to better approximate the variability of the population, which decreases

    the amount of error that is expected between the sample and population.

3. $\mu = 25$; $\sigma_M = \dfrac{4}{\sqrt{12}} = 1.15$. Two SE's above the mean would be a mean of 27.3.

4. $\mu = 32$; $\sigma_M = \dfrac{10}{\sqrt{23}} = 2.09$. 1.5 SE's above the mean would be a mean of 35.13.

5. $\mu = 54$; $\sigma_M = \dfrac{13.4}{\sqrt{8}} = 4.74$. 3.2 SE's below the mean would be a mean of 38.84.

True/False Questions

1. The central limit theorem states that the distribution of sample means will be normally

    distributed, but only if the underlying distribution is skewed.

2. The sampling distribution of the mean is made up of all the possible samples of a particular

    sample size.

3. With a sample size of $n = 1$, the standard error will be equal to the standard deviation.

4. Increasing sample size will decrease the standard error.

5. A score that falls two standard deviations above the mean falls in a different location on the

    normal curve than a mean that falls two standard errors above the mean.

True/False Answers

1. False   3. True   5. False

2. True   4. True

138

Short-Answer Questions

1. What is the central limit theorem? Why is it important to inferential statistics?

2. Why is the mean of the sampling distribution of the mean always equal to the population mean?

3. Where do most individual sample means fall in relation to the population mean? To what statistical concept do we attribute the differences between individual sample means and the population mean?

4. Why does the standard error of the mean shrink as the size of your sample increases?

5. If you draw an infinite number of samples from a negatively skewed population, what will be the shape of the distribution of these sample means? Why does it take this shape?

Answers

1. The central limit theorem states that the distribution that results from drawing an infinite number of samples of the same size from any population will be normal. This is important to inferential statistics because it allows the proportions of the normal curve to be used for hypothesis testing.

2. It is expected that the majority of the sample means will fall close to the actual mean of the population. Therefore, the sample means will be distributed symmetrically around the population mean.

3. We expect sample means to fall close to the population mean, which is why the distribution of sample means is normally distributed. Differences between the population mean and sample mean are typically associated with sampling error.

4. As the sample size increases, the sample is better able to approximate the variability of the population. As the sample becomes a more accurate representation of the population, there is less expected deviation between the samples and the population.

5. The distribution will be normally distributed. This is a function of the central limit theorem, which informs us that the sample means will be normally distributed around the population mean.

Multiple-Choice Questions

1. According to the central limit theorem, if a population is very positively skewed, the resulting distribution that would be created by samples of $n = 45$ will be

   a. Normally distributed

   b. Positively skewed

   c. Negatively skewed

   d. Bi-modal

2. The standard error will almost always be _____ compared to the standard deviation

   a. Larger

   b. Smaller

   c. Equal

   d. It depends on the sample size

3. The standard error for a distribution of samples of size $n = 1$ would be

   a. 1

   b. 2

   c. 3

   d. σ

4. For a population with $\sigma = 6$, what would be the standard error for samples of size 36?

   a. 6

   b. 36

   c. 1

   d. 54

5. For a population with $\sigma = 10$, what would be the standard error for samples of size 45?

   a. 1.49

   b. 2

   c. 3.41

   d. 4.32

Multiple-Choice Answers

   1. A

   2. B

   3. D

   4. C

   5. A

140

Module Quiz

1. A real estate company is reviewing its properties in a major city in America. It owns a substantial number of houses across several low-income and high-income areas. If the company looks at the average of all of its properties, the distribution is slightly positively skewed. Disappointed with this information, the company's researchers break up the properties into smaller groups (samples) based on their location, such that they have 50 groups of 10 properties. If they were to graph the distribution of these means, how would the researchers expect this distribution to appear?

2. The population of property values has a $\mu = 250$ thousand dollars with a $\sigma = 25$ thousand dollars. A wealthy family is looking to move to this city, and the real estate group will show them a sample of $n = 20$ houses that is 2.5 SE's above the mean. What is the average cost of a house in this sample?

3. The real estate company discovers that it can take its condominiums and divide them into smaller condominiums for college students, with the units still averaging $250,000 in value. If it can divide each condominium into 4 condominiums, the number of units per sample increases to $n = 80$. Without doing any calculations, how would you expect the standard error to change based on this adjustment?

4. A group of college students is looking for an inexpensive apartment. Ideally, they would like something that is a little less expensive than the mean. Using the sample of $n = 80$, what is the average cost of a condominium that is 1.2 SE's below the mean? (As above, $\mu = 250$ thousand dollars with a $\sigma = 25$ thousand dollars.)

5. Based on the calculations of the SE you did for question 2 and question 4, which standard error would the family in question 2 and the college students in question 4 prefer when looking for the lowest cost of a house/condominium?

Quiz Answers

1. They would expect it to be appear normally distributed due to the central limit theorem.

2. $\mu = 250$; $\sigma_M = \dfrac{25}{\sqrt{20}} = 5.59$. The mean price of these houses would be 263.98.

3. The standard error is expected to decrease because the sample size has increased.

4. $\mu = 250$; $\sigma_M = \dfrac{25}{\sqrt{80}} = 2.80$. The mean of these condominiums would be 246.65 thousand

dollars.

5. The family in question 2 would prefer housing based on the SE obtained in question 4. The SE in question 4 was substantially smaller, indicating that the actual average cost of a house 2.5 SE's above the mean would be substantially lower. In contrast, the college students in question 4 would have preferred the SE obtained in question 2. They were searching for condominiums 1.2 SE's below the mean and using the larger SE from question 2. They would have found a lower average condominium cost with the larger SE.

Module 16

Learning Objectives

- Distinguish between a *z*-score and a *Z* test

- Know the conditions under which it is appropriate to use a normal deviate *Z* test

- Calculate a normal deviate *Z* test

- Use a normal curve table to interpret a normal deviate *Z*

Module Summary

- The primary purpose of a hypothesis test is to determine if the difference between the sample mean and the population mean is substantial enough to believe that the samples come from a different population (indicating that the IV had an effect). This logic is embodied in the formula for a *Z* test, or a ***normal deviate Z test***. This formula determines the difference between the sample mean and the population mean in the numerator ($M - \mu$) and then compares this to the average amount a sample mean should deviate from the population mean, or the standard error of the mean. Thus, the formula appears as follows:

$$Z_{\text{Norm Dev}} = \frac{M - \mu}{\sigma_M}$$

- The result from this test will tell you how far the sample mean falls from the population mean in standard error units. You can then refer to the table in Appendix A to determine the percentage of obtaining a sample mean above or below the mean of your current sample, just as you would for an individual *z* score. Based on this probability, you can determine the chances that you would have obtained the sample from the population in the comparison.

- In research, the population from which a sample is drawn is rarely known, and this test can help to infer about the population from which the sample was drawn. For example, you

143

obtain a sample mean with a z-score of 2. This means that you would expect to draw a sample with this mean or higher 2.28% of the time from this population. In other words, the chances of drawing a sample with this mean, from this population, are so slim that it is likely that this sample came from a different population.

Computational Exercises

1. Using the following values, compute the normal deviate test: $\mu = 32$, $M = 28$, $\sigma_M = 8$.

2. The principal of a high school is seeking to improve his school's performance on a foreign language test by recruiting a new teacher for one of the classes. Last year, the school had a $\mu = 83$ and $\sigma = 5$. The new class has $n = 32$ students and obtains a mean grade of 87 on this year's test. He wonders if this teacher has been able to improve his students' performance on the test. A) What are the hypotheses? B) Are the chances of obtaining such a mean greater or less than 5%?

3. The owners of a retirement community want to improve the satisfaction of their residents by implementing a new exercise activity. Prior satisfaction, as rated by a 0-10 scale, was $\mu = 6$ and $\sigma = 1.2$. Only $n = 12$ residents participate in the new activity, and they rate their satisfaction as an $M = 8$. The owner wonders if the new activity improved satisfaction ratings. A) What are the hypotheses? B) Are the chances of obtaining such a mean greater or less than 1%?

4. An appliance company is trying to improve the efficiency of its blow-dryer. The old model would dry hair in $\mu = 130$ seconds with a $\sigma = 22$. The new blow-dryer was able to dry a sample of $n = 12$ people's hair in $M = 120$ seconds. What is the probability that one of the old blow-dryers would be able to dry hair as fast or faster than this newer model?

5. A fashion designer wants to determine the effect that changing the color of her headbands would have on their sales. The headbands currently sell $\mu = 8$ units per week, with a $\sigma = 1$,

144

in stores across the nation. In a conservative move, the designer releases the new color headbands to only $n = 4$ stores and monitors their sales for the week. The new headbands sell with an $M = 10$ during the week. A) What are the hypotheses? B) Are the chances of obtaining such a mean greater or less than 5%?

Computational Answers

1. $z = 0.50$.

2. a. Null: The new teacher will not improve the performance of his students. Alternative: The new teacher will improve the performance of his students.

   b. $\sigma_M = \dfrac{5}{\sqrt{32}} = .88$; $z = \dfrac{87 - 83}{0.88} = 4.55$; the chances are less than 5%.

3. a. Null: The new program will not improve resident satisfaction. Alternative: The new program will improve resident satisfaction.

   b. $\sigma_M = \dfrac{1.2}{\sqrt{12}} = 0.35$; $z = \dfrac{8 - 6}{.35} = 5.71$; the chances are less than 1%.

4. $\sigma_M = \dfrac{22}{\sqrt{12}} = 6.36$; $z = \dfrac{120 - 130}{6.36} = -1.57$; there is approximately a 6% chance (directional test) that one of the old blow-dryers would dry hair this quickly or faster.

5. a. Null: The new color will not affect headband sales. Alternative: The new color will affect headband sales.

   b. $\sigma_M = \dfrac{1}{\sqrt{4}} = 0.5$; $z = \dfrac{10 - 8}{.5} = 4$; the chances are less than 5%.

True/False Questions

1. The normal deviate test will tell you if your sample came from a specific population.

2. A directional hypothesis indicate that the sample mean is expected to be either above or below the population mean.

3. In the normal deviate test (Z test), a smaller standard error means that the distribution appears more leptokurtic.

4. The "distance" between a sample mean and the population mean is the only information you need to determine if you should reject or retain the null hypothesis.

5. A smaller standard error will increase the chances that you will reject the null hypothesis.

True/False Answers

1. False   3. True   5. True

2. True   4. False

Short-Answer Questions

1. What pieces of information does a z-score from a normal deviate test provide?

2. What information is necessary in order to complete a normal deviate test?

3. How are statistical decisions made using a normal deviate Z test?

4. Although the normal deviate test provides excellent information, it can sometimes be difficult to use in practice. What about the information required to use this test would make it difficult to use?

5. The life expectancy for a dog of a particular breed is 14 years with a σ of 3. Nathan's 4 dogs have lived for an average of 20 years. How unusual is this longevity for this sample of dogs?

Answers

1. A normal deviate z score provides the distance and the location (above or below) of the sample mean from the population mean.

2. A normal deviate test requires a sample mean, a population mean, a population standard deviation, and the size of the samples used to make the distribution of sample means.

3. First, the distance of the sample mean from the population mean is compared to the standard error, the expected average deviation. Then, the probability of obtaining such a sample mean is determined by using the Z table. If the probability of obtaining this mean is less than the critical value (or value that denotes the allowed Type 1 error level), then the null is retained. However, if the probability of obtaining this ample mean is greater than the critical value, the null is rejected.

4. The normal deviate test requires population information, which is not always readily available in practice.

5. A normal deviate Z test for Nathan's dogs would provide a z score of 4. The odds of finding a sample of 4 dogs who have lived this long is much less than 1%.

Multiple-Choice Questions

1. The grades on a Spanish test in a foreign language class for the past 5 years have shown a $\mu = 75$ with a $\sigma = 4$. This year, the class average was 80 on the test, for a class of $n = 30$. Simon scored 80, which was the class average. What is the probability that someone did better than *Simon* on this test this year?

   a. 0.75

   b. 0.68

   c. 0.32

   d. 0.50

2. Using the information from question 1, what is the probability that the class from another year would do better than *Simon's class*?

   a. 0.10

   b. 0.00

   c. 0.11

   d. 0.31

3. A developmental psychologist is studying motor development by examining the ages in which children are able to take 3 steps. If the average child can take 3 steps at age $\mu = 1.2$

years with a $\sigma = 0.3$ years, what is the probability of finding a group of $n = 5$ children who could walk at age 0.9 years?

    a. 0.16   c. 0.01

    b. 0.99   d. 0.13

4. Using the information from question 3, what is the probability of finding a group of $n = 12$ children who could walk at a mean age of 1.3 years or greater?

    a. 0.30   c. 0.00

    b. 0.01   d. 0.13

5. A guitar typically retails for $\mu = \$200$, $\sigma = \$25$. A popular online auction Web site has the guitar available from $n = 8$ sellers who are selling the guitar for $M = \$189$. If you want to purchase 8 guitars for your music shop, what is the probability that you would find another sample of 8 guitars for less?

    a. 40%   c. 23%

    b. 11%   d. 89%

Multiple-Choice Answers

    1. D   4. D

    2. B   5. B

    3. C

Module Quiz

1. A basketball team named the "Magic Stars" consists of $n = 10$ players. The Magic Stars are attempting to improve their shooting accuracy. Currently they make $M = 0.24$ of their shots in a league where the average team of 10 makes $\mu = .34$, $\sigma = 0.11$, shots. Using the team's current average, would say that they are shooting significantly worse than the league?

2. The Magic Stars hire a new coach who gives them more drills and exercises to improve their shooting. This brings the teams average up to $M = 0.32$. How much lower than the rest of the league are the Magic Stars performing with regard to shooting? Considering your answers to question 1, do you think these new exercises have helped the Magic Stars? ($\mu = .34$, $\sigma = 0.11$, $n = 10$)

3. A new treatment for generalized anxiety disorder is tested by comparing a sample of $n = 40$ individuals' ratings of anxiety to the ratings of the general population ($\mu = 12$, $\sigma = 9$) on a clinician administered rating scale. After the treatments, the sample reports an $M = 16$ anxiety rating. Has the treatment brought anxiety symptoms for this group to within a normal range?

4. Dismayed by the findings in question 3, the developers of the treatment determine the parameters for a different population of individuals with generalized anxiety. The ratings of this population on the measure used to assess anxiety are $\mu = 20$, $\sigma = 8$. With this new information, does the treatment seem effective? ($M = 16$, $n = 40$.)

5. How do probabilities help to guide decision making using the normal deviate test?

Quiz Answers

1. $\sigma_M = \dfrac{0.11}{\sqrt{10}} = 0.03$; $Z = \dfrac{0.24 - 0.34}{0.03} = -2.87$; the Magic Stars are shooting at the lower 0.2% in the league. It is safe to say the Magic Stars are doing significantly worse than the rest of the league.

2. $\sigma_M = \dfrac{0.11}{\sqrt{10}} = 0.03$; $Z = \dfrac{0.32 - 0.34}{0.03} = -0.57$; the Magic Stars are performing at 28% in the league. Considering their change from the 0.2% to the 28% level, it is fair to say that these exercises helped them improve their shooting.

3. $\sigma_M = \dfrac{9}{\sqrt{40}} = 1.42$; $Z = \dfrac{16-12}{1.42} = 2.81$. Only 0.2% of the population is more anxious than

this sample after treatment. Based on this, it seems that the treatment was not effective in

making this sample of people as anxious as the general population.

4. $\sigma_M = \dfrac{8}{\sqrt{40}} = 1.26$; $Z = \dfrac{16-20}{1.26} = -3.16$. Based on this finding, less than 1% of the

population with generalized anxiety has lower anxiety scores. Based on these findings, it

seems that the treatment is effective.

5. A normal deviate test will provide you with the probability that you would obtain such a

sample under the conditions of the population. If the probability is very low, then you can

assume that your sample is a rare occurrence and should reject the null. In contrast, if the

odds of obtaining your sample are high, then it is probably a common occurrence in the

current population and you should retain the null.

Module 17

Learning Objectives

- Distinguish between a normal deviate Z test and a one-sample t test

- Know the conditions under which it is appropriate to use a one-sample t test

- Understand the similar logic underlying various test statistics

- Understand the concept of degrees of freedom

- Distinguish between biased and unbiased estimators of a population parameter

- Find the regions of retention and rejection

- Understand the relationship between directionality of the hypothesis and the tail in the sampling distribution

- Calculate a one-sample t test

- Use a table to interpret calculated t

- Report results in APA format

Module Summary

- A one sample t test is very similar to a Z test in that it follows similar logic and uses a similar process in determining whether or not a sample is significantly different from a population. However, a t test is used in lieu of a Z test when either (1) the size of your sample is less than 25 or (2) you do not know the standard deviation for the population.

- The main distinction between the formula for a one-sample t test and a Z test is the denominator, or the standard error. In a Z test, the standard error is calculated by dividing the population standard deviation by the square root of the sample size (n). In a one-sample t test, the population standard deviation is unknown. Therefore, you must use the ***estimated***

151

population standard deviation, which places $n - 1$ in the denominator. The formula for the estimated population standard deviation is as follows:

$$\sigma_{est} = \sqrt{\frac{\sum(X-M)^2}{n-1}}$$

- The denominator of the estimated population standard deviation is also the degrees of freedom. ***Degrees of freedom*** refer to the number of values that are allowed to vary when calculating a statistic. For example, imagine that you have a sample of size $n = 3$ with $M = 10$. Your first two scores could be any values that you choose, but the third value must make the overall mean of the three scores 10. Let's say that the first two scores were a 7 and a 4. This indicates that in order for the mean of this sample to be 10, the third value must be 19. For this example, there were two degrees of freedom—two scores could have been any value. Thus, the formula for degrees of freedom for a *t*-test is $n - 1$. This formula will change depending on the type of test you are conducting.

- The reason that $n - 1$ is used in the denominator when estimating the standard deviation of a population is to correct for bias. The variability of a sample will always be smaller than or equal to the variability of a population. This is because a sample will tend to miss many of those extreme scores, or outliers, in a population that increase the population variability. Because of this, the standard deviation of a sample is a ***biased indicator*** in that it will almost always *underpredict* (be smaller) than the population standard deviation. To correct for this bias in the sample, we subtract 1 from the denominator to increase the value of the standard deviation, which will better approximate the population standard deviation. This $n - 1$ is called a ***correction factor***. When $n - 1$ is used in the denominator of a standard deviation, that standard deviation is referred to as an ***unbiased indicator***.

- The null hypothesis is retained when the difference between the population mean and the sample mean is *close to zero*. In contrast, the null is rejected when the difference between the population mean and sample is *far from zero*. This is determined by dividing the sampling distribution into two pieces. The piece that represents the area close to zero (or close to the mean) is the *region of retention.* The area that falls far from the mean, or in the tail(s), is referred to as the *region of rejection*.

- In a *nondirectional* test, the alternative hypothesis states that there is a difference between the two samples, but it does not specify in which direction the difference is expected. In other words, there could be a significant increase or a decrease. In contrast, a *directional* test does indicate a specific direction; the sample mean is expected to be higher or lower than the population mean. This is important for determining where the region of retention and rejection are located. For a directional test, the region of rejection is placed entirely in one tail. This type of test is referred to as a *one-tailed* test. For a nondirectional test, the region of rejection is placed in both tails (with half of the α in each). This type of test is referred to as a *two-tailed* test.

- After determining where the region of rejection will fall (one or two tails), you should then decide the level of Type 1 error that you are willing to accept. This refers to the chance that the null may be incorrectly rejected.

- Another aspect of note for the one-sample t-test is the shape of the distribution that is used to determine the probability of obtaining this particular sample mean. In a Z test, the shape of the distribution will always be the same, so the probabilities of that curve area will always be the same. However, because we are using estimates (sample statistics) in our inferential statistics, the shape of the distribution must be amended. We can expect that with smaller sample sizes, the shape of the distribution will deviate further from normality. Thus, the t distributions, because they are actually a family of distributions, have slightly raised tails and are somewhat leptokurtic. However, as the sample size increases past 30, the t distributions that are used are practically identical to the normal curve.

- The scores that mark the region of rejections, or the **critical values**, for the t distributions can be found in Appendix C. The table is used as follows. First, find the degrees of freedom that correspond to the sample size. If the sample size is not listed, then use the next lowest sample size. Then, find the column that corresponds to the error rate. Be mindful that the table provides error rates for both one-tailed and two-tailed tests. Then, find the t value that corresponds to your degrees of freedom (df) and error rate. If the t value you have obtained from your t test exceeds the value in the table, then the null hypothesis is rejected. Conversely, if the t value you have obtained does not exceed the value in the table, the null is retained.

- The one-sample t test is calculated in a manner that is strikingly similar to that of the normal deviate Z test. The primary difference is that the estimated population standard deviation is used to calculate the standard error rather than the population standard deviation. Given this, the formula is very similar to that of a normal deviate Z test.

$$t_{1-samp} = \frac{M - \mu}{\sigma_M}$$

Computational Exercises

1. A farmer is curious about the effect that new plant food will have on her tomato crop. Previously, the tomatoes she grew were $\mu = 0.8$ lbs. She uses the new plant food and grows $n = 13$ tomatoes that weigh an $M = 1.1$ lbs with $\sigma_{est} = 0.7$. Is it likely that she would obtain this crop?

2. You notice an advertisement for a typing course that claims it vastly improves a person's typing rate. It states that the average reading rate is $\mu = 50$ words per minute and a class of $n = 30$ improved to 65 words per minute by using the course. You are skeptical of this advertisement and are able to track down all 30 participants and re-test them. Sure enough, you obtain a sample mean of $M = 65$, but $\sigma_{est} = 41$. Is this course really able to improve typing rate? A) What are your hypotheses? B) Is this a directional or nondirectional test? C) Determine whether you should reject or retain the null at $\alpha = 0.05$.

3. A national anxiety survey finds that individuals with generalized anxiety disorder (GAD) score at approximately $\mu = 13$ on an anxiety measure. You are developing a new treatment to address GAD and implement it in a sample of $n = 6$ participants. You obtain the following values:

13	9
7	13
8	10

Using this data, determine if your treatment effective at reducing anxiety. A) What are your hypotheses? B) Is this a directional or nondirectional test? C) Determine whether you should reject or retain the null at $\alpha = 0.05$.

4. A researcher has developed a treatment for post-traumatic stress disorder (PTSD) that is designed to be briefer than prior interventions. Currently, PTSD treatment lasts for $\mu = 8$ sessions. The new treatment has been used with $n = 4$ participants and has shown results in 5, 6, 5, and 5 sessions. Using this data, determine if this new treatment is significantly briefer than regular treatments. A) What are your hypotheses? B) Is this a directional or nondirectional test? C) Determine whether you should reject or retain the null at $\alpha = 0.01$.

5. A cognitive psychologist is attempting to determine if college-aged students are able to rotate objects in their minds faster than individuals who are middle-aged. Based on prior research,

middle-aged participants can rotate objects in $\mu = 12$ seconds. A sample of $n = 10$ college-aged participants provides you with the following data. A) What are your hypotheses? B) Is this a directional or nondirectional test? C) Determine whether you should reject or retain the null at $\alpha = 0.01$.

8	12
12	13
9	13
6	8
11	8

Computational Answers

1. $\sigma_M = \dfrac{0.7}{\sqrt{13}} = 0.19$; $t = \dfrac{1.1-0.8}{0.19} = 1.58$. The critical table t at nondirectional 5% with 12 df

= 2.1788 and at nondirectional 10% with 12 df = 1.7823. The calculated t does not meet those values, so the null cannot be rejected. It appears likely that she would obtain this crop.

2. a. Null: The course will not improve typing rate. Alternative: The course will improve typing rate.

   b. This is a directional test.

   c. $\sigma_M = \dfrac{41}{\sqrt{30}} = 7.49$; $t = \dfrac{65-50}{7.49} = 2$. Tabled critical t at directional 5% with 29 $df =$

   1.699 and thus we can reject the null. Typing did improve significantly.

3. a. Null: The treatment will not reduce the scores in individuals with GAD. Alternative: The treatment will reduce the scores in individuals with GAD.

   b. This is a directional test.

   c. $\sigma = \sqrt{\dfrac{32}{5}} = 2.53$; $\sigma_M = \dfrac{2.53}{\sqrt{6}} = .42$; $t = \dfrac{10-13}{0.42} = -7.11$.

   Critical table t at directional 5% error with 5 df = 2.02. Calculated t exceeds this value and thus the null is rejected.

4. a. Null: The new treatment will not take fewer sessions to reduce PTSD symptoms.

    Alternative: The new treatment will take fewer sessions to reduce PTSD symptoms.

    b. This is a directional test.

    c. $\sigma = \sqrt{\dfrac{0.75}{3}} = 0.50$; $\sigma_M = \dfrac{0.50}{\sqrt{4}} = 0.25$; $t = \dfrac{5.25-8}{0.25} = -11$. Tabled critical t at

    directional 1% error with 3 $df = -3.1824$. Thus, we can reject the null. The treatment is

    obtaining significant findings in fewer sessions.

5. a. Null: College-aged participants cannot rotate objects faster than middle-aged individuals.

    Alternative: College-aged participants can rotate objects faster than middle-aged individuals.

    b. This is a directional test.

    c. $\sigma = \sqrt{\dfrac{25}{9}} = 2.49$; $\sigma_M = \dfrac{2.49}{\sqrt{10}} = 0.25$; $t = \dfrac{10-12}{0.25} = -8.02$. Tabled critical t at directional

    1% error with 9 $df = -2.622$. The obtained value is greater than the critical value, so the

    null is rejected. College-aged participants rotated objects faster.

True/False Questions

1. The estimated population standard deviation is a biased estimator in that it will always

    overestimate the population standard deviation.

2. The null hypothesis refers to the question "is the difference between the sample mean and the

    population mean close to zero?"

3. The region of retention is the portion of the curve that is in the tails.

4. The t distribution is actually a family of distributions, as opposed to a single distribution.

5. A one-sample t test differs from a Z test in that it uses the estimated population standard

    deviation.

1. False   4. True

2. True   5. True

3. False

Short-Answer Questions

1. When would you use a one-sample *t* test instead of a normal deviate test?

2. Why is the estimated population standard deviation a biased indicator?

3. Compare and contrast the region of rejection for a nondirectional and a directional test.

4. How do the *t* distributions differ from the standard normal distribution? How are the *t* distributions affected by sample size?

5. Why would a researcher move his alpha level from .05 to .01?

Answers

1. A one-sample *t* test is used when the population standard deviation is unknown. Also, some statisticians feel that the *t* test should be used when the sample size is less than 25.

2. The estimated population standard deviation will almost always underestimate the full variability of the population because it is unable to incorporate many extreme scores. To correct this bias, 1 is subtracted from the denominator.

3. In a directional test, the region of rejection is located entirely in one tail. In a nondirectional test, the region of rejection is divided into two parts that are of equal size and located in each of the tails.

4. The *t* distributions are considered to be slightly more leptokurtic than the normal distribution. Also, the tails of the *t* distributions are slightly elevated from the underlying number line. As

sample size increases, the t distributions begin to better mimic the standard normal distribution.

5. This would be done to reduce the chances of making a Type 1 error. This sets up more stringent criteria for rejecting the null.

Multiple-Choice Questions

1. As sample size decreases, how does the t distribution change?

    a. It approaches the normal distribution

    b. It becomes more platykurtic

    c. The tails become more elevated from the number line

    d. B & C

2. Tessa is studying the behavior of beetles. She wants to determine the effect of a new neurochemical on the amount of eggs a female beetle lays. Without the neurochemical, beetles lay $\mu = 14$ eggs. With the neurochemical, the $n = 25$ beetles lay $M = 15.3$ eggs with $\sigma_{est} = 2.5$. What should Tessa conclude about the effect of the neurochemical if she is willing to accept a 5% Type 1 error rate?

    a. Reject the null of a nondirectional test

    b. Retain the null of a directional test

    c. Reject the null of a directional test

    d. Retain the null of a nondirectional test

3. Sari is a huge fan of her college's basketball team. Her team has lost the past few games, and she is determined to figure out the cause of this streak. She hypothesizes that her team has shorter players than the rest of the league. Using the college basketball database, she determines that the μ height is 87 inches. The mean height of her team with $n = 12$ players is 85.3 with $\sigma_{est} = 3.2$. If Sari accepts an error rate of 1%, what should she conclude?

159

a. Reject the null of a nondirectional test

b. Retain the null of a directional test

c. Reject the null of a directional test

d. Retain the null of a nondirectional test

4. A shoe company is trying to improve the walking comfort of its high heels. The results from previous tests have rated the comfort of the heels at a $\mu = 45$ on a scale of 0-100. The company has created a new shoe and gives it to a sample of $n = 8$ to provide comfort ratings. If the new sample has a rating of $M = 50$ with $\sigma_{est} = 16$, what should the company conclude about the new shoes?

a. Reject the null of a nondirectional test

b. Retain the null of a directional test

c. Reject the null of a directional test

d. Retain the null of a nondirectional test

5. A restaurant is trying to improve the amount of time it takes to complete the orders of customers. The average order currently takes $\mu = 15.4$ minutes to complete. The owners try a new incentive plan in which they will pay the waiters a bonus if they can fill an order in less than that time. After a week of monitoring this process, a sample $n = 100$ orders have been filled in $M = 11.3$ minutes with $\sigma_{est} = 6.2$ minutes. What should the owners conclude about this incentive program?

a. Reject the null of a nondirectional test

b. Retain the null of a directional test

c. Reject the null of a directional test

d. Retain the null of a nondirectional test

Multiple-Choice Answers

1. D   3. B   5. C

2. A   4. B

Module Quiz

1. Scientists are seeking to improve the ability an fMRI to detect tumors. Currently, the

    machine has an average success rate of $\mu = .85\%$. Using a sample of $n = 15$ participants that

    were diagnosed with tumors, an updated fMRI machine has a success rate of $M = .91\%$ with

    $\sigma_{est} = .12\%$. Based on these findings, should the scientists conclude that the updated machine

    is better at detecting tumors at the 0.05 level?

2. A hardware store owner wants to improve the sales of lumber over last month's sales of $\mu =$

    120. The owner decides to run a promotion for a month to see if that improves sales. Using

    a sample of $n = 11$ lumber sales, the average of these sales was of $M = \$145$ with $\sigma_{est} = \$20$.

    Based on these findings, should the owners of the star conclude that the promotion increased

    sales at the 0.05 level?

3. Several night clubs are hoping to increase the number of customers during the Memorial Day

    weekend. During the past years, they have had an average of $\mu = 230$ per night for this

    weekend. For this year's Memorial Day holiday, the clubs decide to run an event that will

    increase their notoriety. Over the course of the $n = 3$ days of this weekend, the clubs attract

    $M = 310$ customers with $\sigma_{est} = 45$. Based on these findings, did the event improve patronage

    during the Memorial Day weekend at the 0.05 level?

4. Individuals with social anxiety tend to stare at people with threatening expressions longer

    than people with happy expressions. As a young researcher, you are interested in seeing if

    those with social anxiety are more likely to stare at people with threatening expressions as

    compared to those with no (neutral) expressions. The average length of time a person with

161

social anxiety stares at a person with a threatening expression is $\mu = 3$ seconds. You conduct a study with a sample of $n = 12$ socially anxious individual and discover that they stare at people with neutral expressions for $M = 2$ seconds with $\sigma_{est} = 1.25$ seconds. Using an alpha level of .01, does it seem as though individuals with social anxiety stare at threatening faces longer than they do neutral faces?

5. What is the relation between the size of the critical region for the t distribution and the size of the sample (i.e., the degrees of the freedom)?

Quiz Answers

1. $\sigma_M = \dfrac{0.12}{\sqrt{15}} = 0.03$; $t = \dfrac{0.91 - 0.85}{0.03} = 1.94$. The null hypothesis should be retained. The machine is not better at detecting tumors.

2. $\sigma_M = \dfrac{40}{\sqrt{11}} = 12.06$; $t = \dfrac{145 - 120}{12.06} = 2.07$. The null hypothesis should be retained. The promotion did not substantially improve sales of lumber.

3. $\sigma_M = \dfrac{45}{\sqrt{3}} = 25.98$; $t = \dfrac{310 - 230}{25.98} = 3.08$. The null hypothesis should be rejected. The event improved patronage over previous years.

4. $\sigma_M = \dfrac{1.25}{\sqrt{12}} = 0.36$; $t = \dfrac{2 - 3}{0.36} = -2.77$. The null hypothesis should be rejected. Socially anxious individuals stare at neutral faces for a shorter amount of time than they do threatening faces.

5. Degrees of freedom and the size of the critical region are negatively related. That is, as the degrees of freedom increase, the size of the critical region decreases.

Module 18

Learning Objectives

- Distinguish between tabled and incurred alpha

- Understand the relationship between error and confidence

- Estimate parameters, both point and interval

Module Summary

- In statistics, **confidence** refers to the probability that a Type 1 error was *not* made. In other words, it's the chance that we have found a "real" effect (the changes in our DV are really caused by our IV).

- Using the tables, it is impossible to determine the actual confidence for our hypothesis test. This is because the table does not provide the specific probability, or **p**, of actual incurred error. We can, however, come quite close to actual incurred error by interpolating between table columns. Software such as SPSS calculates actual incurred errors precisely. Using p, confidence can be expressed as $1 - p$.

- It is common to express your results in APA format, which appears as follows for a one-sample t test: $t(df) = t$ test value, $p <$ or $> p$ value; for example, $t(10) = 4.65, p < .05$.

- The decision to reject or retain the null is based on where the test statistic (value obtained from your t test) falls on the t curve, in the region of rejection or retention. Once again, these areas are set by the predetermined α, which is conventionally set at .05. The use of α makes rejecting or retaining the null a dichotomous decision, which can be problematic. The α level that is used can determine the results of a study, which can have larger implications. It may appear unfair that there is no "middle ground" decision in this matter. As such, you should not consider the rejection or retention of the null as a divine message. Rather, you should

163

view it as support for one hypothesis, but be open to the possibility that these results may differ with another sample. However, you can feel more confident in your decision as the *p* value decreases because this means your chances of making a Type 1 error becomes increasingly small.

- *Parameter estimations* are estimations of population parameters that are based on sample statistics. There are two types of parameter estimates. The first is referred to as a point estimate. *Point estimates* are made when using a sample statistic to make a single guess about a population parameter. Stating that the population mean is 7 is a point estimate. The second parameter estimation is an interval estimate. *Interval estimates* suggest that the population parameter falls within a range of values.

- *Confidence intervals* are interval estimates that provide a range of values in which the population parameter is expected to fall. The sizes of these intervals are determined by a set percentage. This percentage refers to the number of times you would expect the population parameter to fall within the interval if the study were done repeatedly. For example, a 75% confidence interval means that 75% of the time, you would expect the population parameter to fall within the interval; 25% of the time, the parameter would not fall within the interval. Confidence intervals are usually set at 95% or 99%. The formula for a confidence interval is as follows:

$$CI = M +/- (t_{crit\ at\ 1/2á})(ó_M)$$

Computational Exercises

1. Previously we met a farmer who was curious about the effect that new plant food will have on her tomato crop. Previously, the tomatoes she grew were $\mu = 0.8$ lbs. She uses the new

plant food and grows $n = 13$ tomatoes that weigh an $M = 1.1$ lbs with $s = 0.7$. Using this information, find a 95% confidence interval for the new crop for this nondirectional hypothesis.

2. Previously, you noticed an advertisement for a typing course that claimed it vastly improved typing rate. It stated that the average typing speed is $\mu = 50$ words per minute and a class of $n = 30$ improved to 65 words per minute. You were skeptical of this advertisement and tracked down all 30 participants and re-tested them. Sure enough, you obtained a sample mean of 65 and a standard deviation (with $n - 1$) of 41. Using this information, find a 99% confidence interval for this directional hypothesis.

3. A national anxiety survey finds that individuals with generalized anxiety disorder (GAD) score at approximately $\mu = 13$ on an anxiety measure. You are developing a new treatment to address GAD and implement it in a sample of $n = 6$ participants. You obtain an $M = 11$ and $\sigma_{est} = 1.3$ from this sample. Using this information, find a 95% confidence interval for this nondirectional test.

4. A researcher has developed a treatment for post-traumatic stress disorder (PTSD) that is designed to be much shorter than prior interventions. Currently, PTSD treatment usually lasts for $\mu = 8$ sessions. The new treatment has been used with $n = 4$ participants and has shown results in $M = 6$, $\sigma_{est} = 2$ sessions. Using this information, find a 95% confidence interval for this directional test.

5. A fashion designer wants to determine the effect that changing the color of her headbands would have on their sales. The headbands currently sell $\mu = 8$ per week in stores across the nation. In a conservative move, the designer releases the new color headbands to only $n = 4$ stores and monitors their sales for the week. The new headbands sell with an $M = 10$ and σ_{est}

= 1 during the week. Using this information, find a 99% confidence interval for this directional hypothesis.

Computational Answers

1. CI = 1.1 +/– (2.179)(0.19) = 1.51 and 0.69.

2. CI = 65 +/– (2.756)(7.49) = 85.642 and 44.358.

3. CI = 11 +/– (2.571)(0.53) = 12.36 and 9.64.

4. CI = 6 +/– (3.182)(1) = 9.18 and 2.82.

5. CI = 10 +/– (5.841)(0.5) = 12.92 and 7.08.

True/False Questions

1. The size of the confidence interval increases as the percentage increases. In other words, a 95% confidence interval is smaller than a 99% confidence interval.

2. Confidence is defined as $1 - p$.

3. α refers to preselected error, whereas p represents actual error.

4. Our decision to retain or reject the null hypothesis is dependent upon our confidence interval rather than our predetermined level of α.

5. Parameter estimation involves using a sample to understand its population of origin.

True/False Answers

1. True

2. False

3. True

4. False

5. True

Short-Answer Questions

1. What is a confidence interval?

2. If a result is written as $p < .05$, can you determine the exact level of confidence? Why or why not?

3. How are confidence intervals affected by the size of the sample?

4. How does the size of the confidence interval change as you decrease the level of confidence (e.g., moving from a 99% to 95% CI)?

5. You have just completed a one-sample t test and discovered that your 99% confidence interval ranges from 8 to 12. If your $\mu = 9$, do you think that you rejected or retained the null hypothesis after completing your t test?

Answers

1. A confidence interval provides a range of values in which the sample mean is likely to fall if the study were replicated numerous times.

2. The exact level of confidence cannot be determined. Confidence reflects the actually probability that a Type 1 error was made. In this situation, the precise p value is unknown; you know only that the chance of making a Type 1 error was less than 5%.

3. Confidence intervals decrease as sample size increases. This is because larger samples will result in a smaller critical value and a smaller standard error. As a result, a larger sample will decrease the standard error, reducing the size of the interval.

4. The size of the confidence interval will decrease as you decrease the level of confidence.

5. The null hypothesis was likely retained because the population mean, or the group to which you were comparing your sample, falls within the confidence interval. If this is the case, you can be confident that your population mean was not different from your sample.

Multiple-Choice Questions

1. What is the 95% confidence interval for a nondirectional hypothesis test in which $M = 13$, $\sigma_{est} = 4$, $n = 36$, and $\alpha = 0.05$?

    a. 17 and 11   c. 14.37 and 12.63

    b. 28.25 and 25.57   d. 7.59 and 10.71

2. What is the 99% confidence interval for a nondirectional hypothesis test in which $M = 24$, $\sigma_{est} = 2.7$, $n = 10$, and $\alpha = .01$?

    a. 27.77 and 21.23   c. 26.7 and 21.3

    b. 37.25 and 30.45   d. 35 and 30

3. Rosa is very nervous about taking the GRE next week. She does a lot of preparation for the test and has taken $n = 7$ practice tests and obtained an $M = 1270$ with $\sigma_{est} = 45$. The population mean for the GRE is 1120. Calm Rosa's nerves by providing her with the 99% confidence interval (nondirectional) for her performance on the practice GRE's, which will help give her an idea of how well she will do on the actual test.

    a. 1315 and 1225   c. 1183.06 and 1056.94

    b. 1165 and 1075   d. 1336.06 and 1206.94

4. What is the 95% confidence interval for a nondirectional hypothesis test in which $M = 23$, $\sigma_{est} = 7$, $n = 12$, and $\alpha = 0.05$?

    a. 25.98 and 23.24   c. 12.32 and 9.32

    b. 27.45 and 18.55   d. 32.47 and 28.82

5. What is the 95% confidence interval for a nondirectional hypothesis test in which $M = 78$, $\sigma_{est} = 6.35$, $n = 21$, and $\alpha = 0.05$?

    a. 95.32 and 87.47   c. 80.89 and 75.11

    b. 65.34 and 62.28   d. 83.44 and 74.35

1. C   4. B

2. A   5. C

3. D

Module Quiz

1. A local college claims to be highly selective in its admissions process. The average SAT score is $\mu = 1000$, and the average SAT score of a person who is admitted to the local college is $M = 1300$, $\sigma_{est} = 200$, $n = 75$. What is the point estimate for the SAT score of a student who is successfully admitted to this college?

2. Using the information for the college mentioned in question 1, what is the 95% (nondirectional) confidence interval for the student who is successfully admitted to this college? ($M = 1300$, $\sigma_{est} = 200$, $n = 75$.)

3. A new yoga studio is opening, and the director wants to obtain an estimate as to how many people will attend on the first week. The national yoga association indicates that first-week attendance is usually $\mu = 250$ people. Within this particular area, $n = 7$ yoga studios reported having $M = 234$, $\sigma_{est} = 34$ people attend during the first week. What is the 95% confidence interval (nondirectional) for the amount of people who will attend the first week?

4. Jamal is training for a marathon and is doing better than expected. He is following a training regime that gradually allows him to run for longer and longer distances. He has done much better than anticipated during his first week of training. For the second week of training, the average distance that a person who is training can run is $\mu = 2.4$ miles. After training for the second week, Jamal is able to run $n = 5$ times for $M = 3.7$ each, with $\sigma_{est} = 1.3$. What is the 99% confidence interval (nondirectional) for Jamal's running the second week?

169

5. A new professor is worried about his ability to gain tenure. The average amount of time it takes a professor to obtain tenure nationwide is $\mu = 6.3$ years. Looking a bit closer at her own institution, it took $n = 12$ professors $M = 5.7$, $\sigma_{est} = 0.67$ years to obtain tenure. What is the 99% confidence interval (nondirectional) for this professor to obtain tenure?

Quiz Answers

1. The point estimate is the mean of the sample, 1300.

2. CI = 1300 +/– (2.042)(23.09) = 1347.16 and 1252.84.

3. CI = 234 +/– (2.447)(12.85) = 265.44 and 202.56.

4. CI = 3.7 +/– (4.032)(0.58) = 6.04 and 1.36.

5. CI = 5.7 +/– (3.106)(0.19) = 6.30 and 5.10.

Module 19

Learning Objectives

- Understand the difference between a sampling distribution of the mean and sampling distribution of the difference between the means

- Understand why the mean of the sampling distribution of the difference between the means is 0

- Understand the impact of sample size on the size of the standard error of the difference between the means

- Understand the impact of the size of the standard error of the difference between the means on the test statistic t

Module Summary

- In two-sample research, we compare two samples that are thought to be representative of separate populations—typically, one treated and another one either not treated or treated differently. This comparison enables us to answer the question, "Are the populations from which these samples are taken likely to be significantly different from one another?"

- The null hypothesis for two-sample research is that there *is not* a difference between the populations or treatments (any difference between the samples is due to sampling error). In contrast, the alternative hypothesis for two-sample research indicates there *is* a difference between the populations or treatments.

- The distribution for two-sample research is called the ***sampling distribution of the difference between the means***. It is created by subtracting the means of all possible sample pairs from the two populations that are being compared. Because the null hypothesis states there is no difference between the differently treated populations, we can assume that the most common mean difference between these pairs of sample will be 0. Also, if the null hypothesis is true, we can assume that there is a low probability of obtaining a very large difference between sample means. This indicates that the sampling distribution of the difference between the means will be normally (symmetrically) distributed with a mean of 0.

- The standard deviation of the sampling distribution of the difference between the means is referred to as the **standard error of the difference between the means (σ_{M-M})**. It is computed by **pooling** (combining) the variance of the two samples that are being compared. When (1) the sample sizes are equal and (2) the samples are independent, you can use the **special-case formula** for the standard error of the difference between the means. This formula is:

$$\sigma_{M_1-M_2} = \sqrt{\sigma_{M1}^2 + \sigma_{M2}^2}$$

- The standard error of the difference between the means is critical for determining if the difference between two samples is a significant or a nonsignificant difference because it will be the denominator of our test statistic. Larger sample sizes will reduce the size of σ_{M-M}, whereas smaller samples sizes will increase the size of σ_{M-M}. A smaller σ_{M-M} will increase the chances of finding a significant difference between two samples, whereas a larger σ_{M-M} will decrease the chances of finding a significant difference.

Computational Exercises

1. What would be the standard error of the difference between the means for two samples in which $\sigma_1 = 7.5$ and $\sigma_2 = 4.27$?

2. What would be the standard error of the difference between the means for two samples in which $\sigma_1 = 6.2$ and $\sigma_2 = 8.36$?

3. What would be the standard error of the difference between the means for two samples in which $\sigma_1 = 9.3$ and $\sigma_2 = 4.98$?

4. What is the standard error of the difference between the means for a study that is comparing the amount of sleep had by a sample of hibernating brown bears with $n = 10$, $\sigma_1 = 12.7$, to another sample of hibernating in black bears with $n = 10$, $\sigma_2 = 5.23$?

5. What is the standard error of the difference between the means for a study that is assessing the efficiency of deep-sea diving tanks in which the tanks of the company "Deep Down Tanks" has $\sigma_1 = 35.23$ and "Undersea Adventure Tanks" has $\sigma_2 = 24.98$?

Computational Answers

1. $\sigma_{M-M} = \sqrt{(7.5^2) + (4.27^2)} = 8.63.$

2. $\sigma_{M-M} = \sqrt{(6.2^2) + (8.36^2)} = 10.41.$

3. $\sigma_{M-M} = \sqrt{(9.3^2) + (4.98^2)} = 10.55.$

4. $\sigma_{M-M} = \sqrt{(12.7^2) + (5.23^2)} = 13.73.$

5. $\sigma_{M-M} = \sqrt{(35.23^2) + (24.98^2)} = 43.19.$

True/False Questions

1. The sampling distribution of the difference between the means consists of the difference between the means of all possible pairs of samples drawn from two populations.

2. There will almost always be a difference between two sample means, even if the null hypothesis is true.

3. The standard error of the difference between the means is obtained by pooling the variances of the two samples and taking the square root of that pooled variance.

4. A two-sample *t* test would be appropriate if you were comparing a sample of SAT scores from a single school to the national SAT average.

5. In order to use the special-case formula for calculating the standard error of the difference of the means, the two samples that you are comparing must have the same sample size.

True/False Answers

1. True
2. True
3. True
4. False
5. True

Short-Answer Questions

1. How do the sampling distribution of the mean and the sampling distribution of the difference between the means differ in their creation? How does this influence the meaning of the standard deviation of the sampling distribution of the difference between the means?

2. What factor influences the size of the standard error of the difference between the means? What two factors influence this factor?

3. What is actually being compared in a two-sample hypothesis test? What do these elements represent? (In other words, what are they placeholders for?)

4. In a two-sample hypothesis test, is it appropriate to expect that the sample means will always be equal if there is no difference in the two population means? If there is a difference between the sample means, what are the two possible causes of this difference?

5. What process is used to develop the standard error of the difference between the means for two samples?

Answers

1. The sampling distribution of the mean is created by plotting all possible sample means from a population. The sampling distribution of the difference between the means is created by plotting the differences between all possible paired samples from the two populations that are

being compared. The standard deviation for this distribution represents the average amount of variation expected from the difference in sample means.

2. The standard error of the difference between the means is influenced by the size of the standard error of the mean for each sample (two samples). The size of each of these standard errors is influenced by the sample size and standard deviation of each sample.

3. In two-sample hypothesis tests, two samples are actually compared to one another. These samples are assumed to be representative of the populations from which they originate, and the samples are placeholders for the actual populations. As such, two-sample tests are considered to be a method to compare two populations for which you do not have any population parameters and instead have only statistics from the two samples.

4. It is appropriate to expect that there will be differences between the means of two samples, regardless of which hypothesis (the null or alternative) is supported. This difference may be caused by (1) a significant difference in the populations that the samples are thought to represent (that is, the treatments) or (2) sampling error.

5. Combine the two standard errors to form the standard error of the difference between the means. This is referred to as pooling.

Multiple-Choice Questions

1. What is meant by pooling the variance?

   a. Combining the variances of two samples to get a single variance

   b. Combining the standard deviations of two samples to get a single variance

   c. Dividing the variance of two groups in two equal portions

   d. Combining the sums of squares of two samples to get a single sum of squares

2. A study is comparing ride times of a sample of people taking public transportation in the morning as opposed to the evening. What is being compared in the hypothesis test?

a. The means of each sample

b. The means of each population

c. The standard deviations of each sample

d. The standard deviations of each population

3. The mean of the sampling distribution of the difference between the means is always equal to
___?

    a. 1   c. 3

    b. 2   d. 0

4. Assuming that the null hypothesis is true means that there is a low probability that the difference between two samples will be very different from zero. How does this affect the sampling distribution of the difference between the means?

    a. It indicates the distribution will positively skewed

    b. It indicates the distribution will be negatively skewed

    c. It indicates the distribution will be symmetrical

    d. It indicates the distribution will be bi-modal

5. In a study comparing the distances pigeons can fly, group 1 has $\sigma_M^2 = 8$. Which value for σ_M^2 for group 2 would lead to the smallest σ_{M-M}?

    a. 25   c. 3

    b. 12   d. 7

Multiple-Choice Answers

1. A   4. C

2. B   5. C

3. D

Module Quiz

1. When conducting a two-sample hypothesis test with the desire to reject the null, is it preferable to have larger standard errors for each sample or smaller standard errors?

2. The formula for the standard error of the difference between the means in this module has been simplified to $\sigma_{M_1-M_2} = \sqrt{\sigma_{M1}^2 + \sigma_{M2}^2}$. What special criterion is required to use this particular formula?

3. What is the standard error of the difference between the means for a comparison between sales at two shoe boutiques if both stores had an equal amount of sales, with Store 1 having $\sigma_1 = 3.2$ and Store 2 having $\sigma_2 = 4.41$?

4. What is the standard error of the difference between the means for a comparison between the distances that two baseball players hit home runs if Dominic has $\sigma_1 = 126.34$ and Rashid has $\sigma_2 = 154.3$?

5. What is the standard error of the difference between the means for a comparison between the refresh rate of two brands of LCD monitors if Brand A has $\sigma_1 = 45.89$ and Brand B has $\sigma_2 = 65.78$?

Quiz Answers

1. Smaller standard errors are preferable because they require less distance between the sample means in order to reject the null.

2. In order to use this formula, the two samples must have the same sample size.

3. $\sigma_{M-M} = \sqrt{(3.2^2) + (4.41^2)} = 5.45$.

4. $\sigma_{M-M} = \sqrt{(126.34^2) + (154.3^2)} = 199.42$.

5. $\sigma_{M-M} = \sqrt{(45.89^2) + (65.78^2)} = 80.21$.

Module 20

Learning Objectives

- Understand the similar logic underlying various test statistics

- Determine the degrees of freedom

- Calculate a two-sample t test for independent samples and equal sample sizes

- Use a table to interpret the calculated t

- Report results in APA format

Module Summary

- The format of the two-sample t test is similar to that of the other hypothesis tests that we have covered. It compares "what you got" (the difference between two sample means; $M_1 - M_2$) to "what you expected" (the expected between difference between the two population means; $\mu_1 - \mu_2$), divided by the "standardized random error" (the standard error of the difference between the means; $\sigma_{M - M}$).

$$t_{2-samp} = \frac{(M_1 - M_2) - (\mu_1 - \mu_2)}{\sigma_{M-M}}$$

- The expected difference between the two populations ($\mu_1 - \mu_2$) under the null hypothesis is always expected to be zero. This reduces the formula to:

$$t_{2-samp} = \frac{(M_1 - M_2)}{\sigma_{M-M}}.$$

- Appendix C contains the t table used to determine if the difference between the sample means is *only a little* (attributed to sampling error) or *a lot* (attributed to the independent variable). As in previous hypothesis tests, the means are considered significantly different if the value obtained from the t test is greater than the value found in the table (the critical t).

The critical t is obtained by using the degrees of freedom, which is found by using the following formula: $df = (n - 1) + (n - 1)$. This formula could also be rewritten as $df = N - 2$ ($N = 2n$).

- After conducting any two-sample t test, the results are usually reported in APA format; t(degrees of freedom) = $t_{obtained}$ value, $p <$ or $> \alpha$. For example, a $t_{obtained} = 3.5$, with 4 degrees of freedom, that is significant at $\alpha = .05$ would be written as $t(4) = 3.5, p < .05$.

Computational Exercises

1. You are interested in comparing the levels of stress among graduate students in a psychology program to the stress among graduate students in a business program. You give them a scale with range of 0 (no stress) to 7 (extremely stressed) and obtain the following data:

Psychology	Business
1	6
5	6
0	2
4	3
2	1
7	7

   a. What are the independent and dependent variables for this study?

   b. State the null hypothesis and research hypothesis.

   c. Can you reject the null hypothesis at $\alpha = .05$?

2. A computer company is interested in seeing if a larger monitor size *increases* the amount of time someone spends on the computer per day. Its researchers give a sample of individuals a 24" monitor and another sample of different people a 15" monitor and ask them record how many hours per day they use their computer. They obtain the following scores:

24"	15"
7	1
8	2
11	3

179

9	4
6	2
7	3
8	3
5	3
9	2

a. What are the independent and dependent variables for this study?

b. State the null hypothesis and research hypothesis.

c. Can you reject the null hypothesis at $\alpha = .05$?

3. College A recently beat College B in a weightlifting competition. The results of the match were very close, with College A's $n = 6$ team members lifting $M = 130$ pounds with $\sigma_{est} = 6$ and College B's $n = 6$ team members lifting $M = 128$ pounds with $\sigma_{est} = 7$. The coach of College B's team would like to use some statistics to help improve his team's spirits. What could he tell them using $\alpha = .05$?

4. Two acting agencies are trying to secure the same high-profile client. The client requests data on how many acting jobs each agency has been able to provide for $n = 15$ of its clients. Agency A provides the following information: $M = 21$, $\sigma_{est} = 3.2$. Agency B provides the following information: $M = 18$, $\sigma_{est} = 4.1$. Is Agency A's record significantly better than Agency B's?

5. Ms. Rose is involved in a law suit and wants to hire a lawyer who will be able to resolve the conflict quickly. She find two law firms, "Turtle & Associates" and "Hare & Associates." If the following information about the length of the average case for $n = 10$ cases for each firm are as follows, which firm should Ms. Rose choose? Turtle & Associates presents data of $M = 4.32$, $\sigma_{est} = 1.1$; Hare & Associates presents $M = 3.87$, $\sigma_{est} = 2.3$.

Computational Answers

1a. Independent variable: graduate program (psychology/business)

Dependent variable: level of stress

1b. Null: There is no difference in the level of stress between the psychology graduate students and the business graduate students.

Research: There is a difference in the level of stress between the psychology graduate students and the business graduate students.

1c. Psychology: $M = 3.17$, SS $= 34.83$, $\sigma^2 = 6.97$, $\sigma = 2.64$, $\sigma_M = 1.08$.

Business: $M = 4.17$, SS $= 30.82$, $\sigma^2 = 6.17$, $\sigma = 2.48$, $\sigma_M = 1.01$.

$$\sigma_{M-M} = \sqrt{1.08^2 + 1.01^2} = 1.48$$

$$t_{2-samp} = \frac{3.17 - 4.17}{1.48} = -0.68$$
$$t_{obtained} = -0.68$$

$$df = 10$$

$t_{critical} > t_{obtained}$, so retain the null.

There is no significant difference between the stress levels of students in the business graduate program and those in the psychology graduate program.

2a. Independent variable: Monitor size

Dependent variable: Hours of computer use

2b. Null: The amount of computer usage by those with a 24" monitor does not differ from that of those with a 15" monitor.

Alternative: The amount of computer usage by those with a 24" monitor is greater than that of those with a 15" monitor.

2c. 24": $M = 7.78$; SS $= 570$; $\sigma^2 = 3.19$, $\sigma = 1.79$, $\sigma_M = 0.60$

15": $M = 2.56$; SS $= 65$; $\sigma^2 = 0.78$, $\sigma = 0.88$, $\sigma_M = 0.29$

$$\sigma_{M-M} = \sqrt{0.60^2 + 0.29^2} = 0.66$$

$$t_{2-samp} = \frac{(7.78 - 2.56)}{0.66} = 7.86$$

$df = 16$

$t_{critical} < t_{obtained}$, so reject the null.

Those with a larger monitor use the computer significantly more than those with a smaller monitor.

3. Team A: $\sigma = 6$, $\sigma_M = 2.45$

   Team B: $\sigma = 7$, $\sigma_M = 2.86$

   $$\sigma_{M-M} = \sqrt{(2.45^2) + (2.86^2)} = 3.76$$

   $$t_{2-samp} = \frac{(130 - 128)}{3.76} = 0.53$$

   The coach could tell his team that although they lost, they did not lose by a statistically significant amount. In other words, the two teams lifted amounts of weight that were comparable, within the bounds of sampling error.

4. Agency A: $\sigma_{est} = 3.2$, $\sigma_M = 0.83$

   Agency B: $\sigma_{est} = 4.1$, $\sigma_M = 1.06$

   $$\sigma_{M-M} = \sqrt{(0.83^2) + (1.06^2)} = 1.34$$

   $$t_{2-samp} = \frac{(21 - 18)}{1.34} = 2.23, p < .05$$

   Yes, Agency A's record is significantly better than Agency B's.

5. Turtle & Associates: $\sigma = 1.1$, $\sigma_M = 0.35$

   Hare & Associates: $\sigma = 2.3$, $\sigma_M = 0.73$

$$\sigma_{M-M} = \sqrt{(0.35^2) + (0.73^2)} = 0.81$$

$$t_{2-samp} = \frac{4.32 - 3.87}{0.81} = 0.56, p > .10$$

Both firms seem to have comparable trial lengths. Ms. Rose could select either firm.

True/False Questions

1. You conduct a study on a new treatment for anxiety. Your results indicate that participants who underwent this new treatment did not have significantly lower anxiety as compared to those who did not undergo the treatment, $t(23) = 1.99, p > .05$. This means the treatment failed in reducing anxiety.

2. In an independent samples t test, the degrees of freedom are $n - 1 + n - 1$.

3. The numerator of the two-sample t test formula compares the obtained difference between the samples to the expected difference between the populations.

4. The numerator for an independent sample t-test should always include the difference between the population means because they are sometimes not equal to 0.

5. Independent sample t tests are always nondirectional.

True/False Answers

1. False   3. True   5. False

2. True   4. False

Short-Answer Questions

1. Researchers are interested in examining the effects of a new medication to treat cancer. To do so, they will conduct a research study in which they will compare the cancer cells in a sample that has received the medication to a sample that has not. What are the null and alternative hypotheses for the study?

2. Why is the formula for a two-sample t test reduced to $t_{2-samp} = \dfrac{(M_1 - M_2)}{\sigma_{M-M}}$ from

$$t_{2-samp} = \dfrac{(M_1 - M_2) - (\mu_1 - \mu_2)}{\sigma_{M-M}}?$$

3. What do each of the three bolded items in the following statement tell you? $t(\mathbf{12}) = \mathbf{4.1}$, $\mathbf{p < .05}$.

4. How does a normal deviate Z test differ from a two-sample t test?

5. If you are provided with the means, sample sizes, and standard deviations for two samples that are going to be used for a two-sample t test, what steps are required to obtain a t value?

Answers

1. Null: There is no difference between the population treated with the medication and the population that did not receive the medication ($\mu_{treated} = \mu_{tunreated}$)

   Alternative: There is a difference between the population treated with the medication and the population that did not receive the medication ($\mu_{treated} \neq \mu_{tunreated}$)

2. The second term in the numerator represents the expected difference between the populations under the null hypothesis. This difference is always expected to be zero, so this term will always be zero. As a result, the formula can be simplified.

3. 12 = the degrees of freedom used to find the critical value

   4.1 = the obtained t value

   $p < .05$ = the t value is significant when $\alpha = .05$

4. A normal deviate Z test compares sample statistics for a single sample to a known population parameter. A two-sample t test compares the difference between two sample means to the expected difference between two populations that would be equal under the null hypothesis. In other words, a normal deviate Z test compares a sample and a population, and a two-

sample *t* test compares the difference in the means of two samples to the difference between the means of two (potentially equal) populations.

5. First, calculate the standard error for each sample. Second, calculate the pooled variance for the two samples. Third, divide the difference between the means by the pooled variance to obtain your *t* value.

Multiple-Choice Questions

1. What is the being compared in the numerator of a two-sample *t* test? (Hint: Refer to the formula on page 233 of the text.)

    a. The difference between the two sample means

    b. The difference between the two population means

    c. The difference between the two sample means and the expected difference between the two population means

    d. The difference between the two samples' variability

2. You are conducting an independent two-sample *t* test with an $N = 30$. What are the degrees of freedom for your study?

    a. 30   c. 15

    b. 28   d. 13

3. Which of the following results is correct and reported in APA format?

    a. $t(13) = 3.45, p < .05$ c. $t = 3.45, p > .05$

    b. $t(13) = 3.45, p > .05$ d. $t = 3.45$

4. Which of the following values would allow to you reject the null hypotheses using a two-sample *t* test with 13 degrees of freedom at $\alpha = .05$, nondirectional test?

    a. 2.10   c. 1.78

    b. 2.45   d. 1.97

5. What is the critical value for a two-sample, nondirectional t test at $\alpha = .01$ with 17 degrees of freedom?

   a. 2.01

   b. 2.90

   c. 1.74

   d. 2.11

Multiple-Choice Answers

   1. C

   2. B

   3. A

   4. B

   5. B

Module Quiz

1. You are conducting a study on a headache relief medicine. You are interested in determining if the medicine provides faster headache relief than aspirin. A sample of $n = 10$ receives the new medication, and another sample of the same size receives aspirin. Both samples report how long it takes them in seconds until they experience pain relief. You obtain the following data: $M_{\text{new medication}} = 64$, $\sigma^2_{\text{est}} = 3$; $M_{\text{aspirin}} = 81$, $\sigma^2_{\text{est}} = 13$. What is the t value?

2. A researcher is examining the influence of a growth hormone on rats. Previous research on this topic provides the population mean and standard deviation for the growth rate of this species of rat. Which hypothesis test would be most appropriate for this experiment?

3. An airline company is seeking a more fuel-efficient plane. Its executives are shown a new prototype that across $n = 14$ flights used an $M = 64$, $\sigma = 3.5$ gallons of fuel per flight.

Compared to their current planes that across $n = 14$ flights use $M = 79$, $\sigma = 2.9$ gallons per flight, should the airline company purchase the new prototypes to save on fuel?

4. Clinical psychologists are developing a new intervention to reduce the frequency of binge eating. They give the intervention to a sample of $n = 12$ people who have binge eating disorder and give an older treatment to another sample of $n = 12$ people with binge eating disorder. They find that those who received the new intervention had $M = 3.2$ and $\sigma = 0.69$ binges. Those who received the older treatment had $M = 3.8$ and $\sigma = 0.21$ binges. Based on this information, does it seem that the new intervention is more effective than the old?

5. Hank and 19 friends ($n = 20$) are going to pursue rock climbing as a form of exercise. Hank joins the local rock climbing organization with the hope of being able to climb walls as quickly as they do. Across $n = 20$ current members, the club averages $M = 5.32$, $\sigma = 1.69$ minutes. After spending two weeks with the club, Hank and his friends each climb an average of $M = 4.30$, $\sigma = 1.78$ minutes. Are Hank and his friends climbing at comparable speed to that of the existing climbing club members?

Quiz Answers

1. $t = -4.029$.

2. A normal deviate Z test.

3. Prototype: $\sigma = 3.5$, $\sigma_M = 0.94$

Older: $\sigma = 2.9$, $\sigma_M = 0.78$

$$\sigma_{M-M} = \sqrt{(0.94^2) + (0.78^2)} = 1.22$$

$$t_{2-samp} = \frac{(64 - 79)}{1.22} = -12.35$$

The prototype planes use significantly less fuel. The company should switch to the new prototype.

4. Newer: $\sigma = 0.69$, $\sigma_M = 0.20$

   Older: $\sigma = 0.21$, $\sigma_M = 0.06$

   $$\sigma_{M-M} = \sqrt{(0.20^2) + (0.06^2)} = 0.21$$

   $$t_{2-samp} = \frac{(3.2 - 3.8)}{0.21} = -2.88$$

   The new intervention is more successful at reducing binge eating than the old intervention.

5. Hank and friends: $\sigma = 1.69$, $\sigma_M = 0.38$

   Existing club members: $\sigma = 1.78$, $\sigma_M = 0.40$

   $$\sigma_{M-M} = \sqrt{(0.38^2) + (0.4^2)} = 0.55$$

   $$t_{2-samp} = \frac{(5.32 - 4.30)}{0.55} = 1.86$$

   There was no significant difference between the times of Hank and his friends and the times of existing members of the club. Hank and his friends are climbing at comparable rates to those of the existing club members.

Module 21

Learning Objectives

- Determine the degrees of freedom

- Perform a two-sample t test for unequal sample sizes

- Use a table to interpret calculated t

- Report results in APA format

Module Summary

- When comparing samples that have unequal sample sizes and are independent of each other, you must use the **generalized formula** for $\sigma_{M} - \sigma_{M}$. This formula uses a process called **weighting** to account for the difference in sample sizes when pooling the variances of samples with unequal n's. The generalized formula is:

$$\sigma_{M-M} = \sqrt{\left(\frac{SS_1 + SS_2}{n_1 + n_2 - 2}\right)\left(\frac{1}{n_1} + \frac{1}{n_2}\right)}$$

- The formula for a t test with unequal sample sizes then becomes:

$$t_{2-samp} = \frac{M_1 - M_2}{\sqrt{\left(\frac{SS_1 + SS_2}{n_1 + n_2 - 2}\right)\left(\frac{1}{n_1} + \frac{1}{n_2}\right)}}$$

- The method we use to determine if the samples are significantly different is the same as that for a t test with equal sample sizes. The degrees of freedom are calculated with the following formula: $df = (n_1 - 1) + (n_2 - 1)$.

Computational Exercises

1. Dr. Smith and Dr. Martinez are comparing the success rates of their different treatments for lung cancer. They are examining success as defined by how long it takes before the patient no longer has lung cancer. Here are their data:

Dr. Smith	Dr. Martinez
$n = 7$	$n = 9$
$M = 35.27$	$M = 32.47$
SS $= 167$	SS $= 178$

   a. State the null hypothesis and the research hypothesis.

   b. Can you reject the null hypothesis at $\alpha = .05$? Report your findings in APA format.

2. You notice that fruit bats and vampire bats are able to fly for different lengths of time. You hypothesize that this difference is attributed to a possible difference in their wing span. You conduct a study to test this hypothesis and obtain the following data.

Fruit Bats	Vampire Bats
$n = 4$	$n = 8$
$M = 8.27$	$M = 10.47$
SS $= 217$	SS $= 258$

   a. What type of t test should be used?

   b. State the null hypothesis and the research hypothesis.

   c. Can you reject the null hypothesis at $\alpha = .05$? Report your findings in APA format.

3. Dr. Johnson is interested in determining if his popularity with students is related to the time of his class. He wants to compare how much his students like him, as assessed by a 7-point rating scale (anchored at 1 = Dislike and 7 = Like), in his 8:30 A.M. class and his 1:00 P.M. class. However, due to the very early time, there are fewer students in his 8:30 A.M. class. He obtains the following data:

8:30 A.M.	1:00 P.M.
$n = 16$	$n = 24$
$M = 3.4$	$M = 5.7$
SS $= 89$	SS $= 95$

190

a. What are the independent and dependent variables for this study?

b. State the null hypothesis and the research hypothesis.

c. Can you reject the null hypothesis at $\alpha = .05$? Report your findings in APA format.

4. Peggy is thinking about hiring a second staff person at her smoothie store in the campus cafeteria. She is trying to determine at what time of day she would need the second person to work, the morning or evening. She observes the number of customers that she receives at the store each morning and afternoon for 10 days, but she loses some of the data for the morning. She is left with the following information.

Morning	Afternoon	
42	72	78
44	83	74
38	74	72
50	78	
37	82	
39	69	
51	83	

a. State the null hypothesis and the research hypothesis.

b. Can you reject the null hypothesis at $\alpha = .05$? Report your findings in APA format.

5. Ken is a dog trainer who wants to see if the obedience class he teaches at his center is able to improve the behavior of dogs more than does his in-home training of dogs. He asks the owners of the dogs that took his obedience class to let him know how many behavior incidents have happened in the month following their completion of his course. He asks the same of those that have received his in-home training. He obtains the following data from his survey.

What should Ken conclude at $\alpha = .01$?

Obedience Class					In-Home Training			
9	10	6	8	7	5	9	10	5
10	10	11	10	8	3	7	3	
9	9	9	11	11	4	5	8	
7	8	10	7		3	7	11	
6	6	9	10		11	8	9	
10	6	10	11		10	7	7	

Computational Answers

1a. Null: There is no difference in the success rate of Dr. Smith and Dr. Martinez

Alternative: There is a difference in the success rates of Dr. Smith and Dr. Martinez

1b.
$$\sigma_{M-M} = \sqrt{\left(\frac{167+178}{7+9-2}\right)\left(\frac{1}{7}+\frac{1}{9}\right)} = 2.50$$

$$t_{2-samp} = \frac{(35.27-32.47)}{2.5} = 1.12$$

1c. Retain the null. Dr. Smith and Dr. Martinez have comparable success rates in their treatment of lung cancer, $t(14) = 1.12$, $p > .05$.

2a. The two-sample t test for unequal sample sizes should be used.

2b. Null: There is no relation between wing size and length of flight time

Alternative: There is a relation between wing size and length of flight time

2c.
$$\sigma_{M-M} = \sqrt{\left(\frac{217+258}{4+8-2}\right)\left(\frac{1}{4}+\frac{1}{8}\right)} = 4.22$$

$$t_{2-samp} = \frac{(8.27-10.47)}{4.22} = -.52$$

Retain the null. It appears there is no relation between wing size and length of flight time, $t(10) = -.52$, $p > .05$.

3a. Independent Variable: Class time (8:30/1:00)

Dependent Variable: Approval rating

3b. Null: There is no difference in Dr. Johnson's approval ratings between his two classes

Alternative: There is a difference in Dr. Johnson's approval ratings between his two classes

3c. $\sigma_{M-M} = \sqrt{\left(\dfrac{89+95}{16+24-2}\right)\left(\dfrac{1}{16}+\dfrac{1}{24}\right)} = 0.71$

$t_{2-samp} = \dfrac{3.4-5.7}{0.71} = -3.23$

The 1:00 P.M. class likes Dr. Johnson more than the 8:30 A.M. class, $t(38) = 3.29$, $p < .01$.

4a. Null: There is no difference in customers between the morning and afternoon shifts.

Alternative: There is difference in customers between the morning and afternoon shifts.

4b. $\sigma_{M-M} = \sqrt{\left(\dfrac{192+8424}{7+10-2}\right)\left(\dfrac{1}{7}+\dfrac{1}{10}\right)} = 2.61$

$t_{2-samp} = \dfrac{43-76.5}{2.61} = -12.84$

The afternoon shift has more customers and needs more assistance, $t(15) = -12.84$, $p < .05$.

5. $\sigma_{M-M} = \sqrt{\left(\dfrac{17.38+77.40}{27+19-2}\right)\left(\dfrac{1}{27}+\dfrac{1}{19}\right)} = 0.44$

$t_{2-samp} = \dfrac{8.81-6.95}{0.44} = 4.25$

Ken should conclude that his obedience class is more effective than his in-home training,

$t(44) = 4.25$, $p < .01$.

True/False Questions

1. In real research, it is common for sample sizes of two independent samples to be equal to one another.

2. The generalized formula for the standard error of the difference of the means provides the same answer as the special case formula when the sample sizes of the two independent samples are equal.

3. Weighting is a process by which the variances of each sample are adjusted based on the *n* of each sample before being pooled.

4. When finding the critical value, a *t* test conducted with unequal sample sizes requires you to consult a different *t* table from the one used for a *t* test with equal sample sizes.

5. When reporting your findings in APA format for a *t* test with independent samples, you should report two different degrees of freedom, one for each sample.

True/False Answers

1. False   3. True   5. False

2. True   4. False

Short-Answer Questions

1. When can you use the generalized formula for the standard error, and when can you use the special case formula?

2. What does weighting do when calculating the standard error for two independent samples with different *n*'s?

3. How is weighting achieved in the generalized formula for the standard error?

Answers

1. The generalized formula can always be used, but it must be used when sample sizes across both samples are not equal. The special case formula is used when the two samples have the same sample size.

2. Weighting combines the variances of two samples proportional to the number of subjects in each sample. In other words, larger samples have a bigger influence on the pooled variance because they are considered to be more representative of the population from which they come.

3. Weighting is achieved by dividing each SS by the n of its own sample.

Multiple-Choice Questions

1. What are the degrees of freedom for a two-sample t test with samples of $n = 10$ and $n = 5$?

    a. 15   c. 10

    b. 13   d. 5

2. What are the degrees of freedom for a two-sample t test that has a sample with an $n = 17$ and another a sample of $n = 21$

    a. 17   c. 38

    b. 36   d. 15

3. What is the standard error for a t test in which one sample has an $n = 32$, SS = 270 and another sample has an $n = 37$, SS = 346?

    a. 2.34   c. 0.73

    b. 1.95   d. 5.14

4. What is the standard error for a t test in which one sample has an $n = 9$, SS = 661 and another sample has an $n = 4$, SS = 693?

    a. 4.80   c. 1.42

    b. 7.02   d. 6.67

5. What is the t value for a t test in which one sample has $M = 34$, $n = 47$, SS = 72,188 and another sample has $M = 66$, $n = 34$, SS = 69,823?

    a. −3.35   b. −1.99

c. −5.10   d. −4.46

Multiple-Choice Answers
Multiple-Choice Answers
   1. B   4. D

   2. B   5. A

   3. C

Module Quiz

1. Tami and Shandi work at the same hair salon and have started to argue about who receives

    bigger tips. Below are their data from the customers they saw last week. Does one seem to

    make significantly more in tips than the other? ($\alpha = .05$)

Tami	Shandi
$n = 45$	$N = 65$
$M = 21.70$	$M = 26.89$
SS = 591	SS = 387

2. Based upon the findings in question 1, Tami decides to find some wealthier clients in an

    attempt to increase her tips. After another week in which she has now seen her wealthier

    clientele, Tami and Shandi compare the following data. Does Tami make significantly more

    in tips than Shandi now? ($\alpha = .05$)

Tami	Shandi
$n = 58$	$n = 65$
$M = 29.35$	$M = 26.89$
SS = 674	SS = 387

3. Matt and Brian are roommates who also work in the same location. They are trying to find

    the fastest way to work in the morning and decide to each take a different route in the

    morning for the week and see how who makes it to work faster. During the week, Brian is

    called into the office on Saturday and Sunday and so commutes two more days than Matt.

    Based on the following data (in minutes), whose route seems optimal for getting into work

    faster? ($\alpha = .05$)

196

	Matt	Brian
	N = 5	N = 7
	M = 60.84	M = 42.86
	SS = 721	SS = 866

4. What potential confound was introduced in the description of the study mentioned in question 3 that may have drastically altered the results in Brian's favor? (HINT: What is different about the extra days that Brian had to drive in to work?)

5. Brian and Matt decide to correct the confound and look at only the 5 days of the work week. Using the following data, whose route seems quicker?

Matt	Brian
60	56
57	60
59	51
66	65
55	60

Quiz Answers

1. $\sigma_{M-M} = \sqrt{\left(\dfrac{591+387}{45+65-2}\right)\left(\dfrac{1}{45}+\dfrac{1}{65}\right)} = 0.58$

$t_{2-samp} = \dfrac{(21.70-26.89)}{0.58} = -8.89$

There is a significant difference between the two, with Shandi making more in tips than Tami, $t(108) = -8.89, p < .05$.

2. $\sigma_{M-M} = \sqrt{\left(\dfrac{674+387}{58+65-2}\right)\left(\dfrac{1}{58}+\dfrac{1}{65}\right)} = 0.54$

$t_{2-samp} = \dfrac{(29.35-26.89)}{0.54} = 4.60$

There is a significant difference between the two, with Tami making more in tips than Shandi, $t(121) = 4.60, p < .05$.

3. $\sigma_{M-M} = \sqrt{\left(\dfrac{721+866}{5+7-2}\right)\left(\dfrac{1}{5}+\dfrac{1}{7}\right)} = 7.38$

$t_{2-samp} = \dfrac{(60.84-42.86)}{7.38} = 2.44$

Brian's route was significantly faster for traveling to work, $t(10) = 2.44$, $p < .05$.

4. Brian had to drive in to work on the weekends, when there is likely to be substantially less traffic. This may have allowed him to arrive at work more quickly and reduced his mean commute time for the week.

5. $\sigma_{M-M} = \sqrt{\left(\dfrac{69.2+109.2}{5+5-2}\right)\left(\dfrac{1}{5}+\dfrac{1}{5}\right)} = 2.99$

$t_{2-samp} = \dfrac{(59.4-58.4)}{2.99} = 0.33$

Based on this data, it appears that there is no difference in the commute time between these routes, $t(8) = 0.33$, $p > .05$.

Module 22

Learning Objectives

- Distinguish between independent and related samples

- Determine the degrees of freedom

- Calculate a two-sample t test for related samples

- Use a table to interpret calculated t

- Report results in APA format

Module Summary

- When the subjects in two samples are not independent, the study is called a ***related-samples*** study. There are two types of related-samples studies. The first is called ***repeated measures***, and it uses the same participants in both samples. The second is called ***matched samples***, and it uses different participants in each sample but matches the participants in each sample on an extraneous variable. The benefit of a related-samples study is that it helps to reduce the influence of the confounding variables that occur when you compare samples containing different people. Examples of these confounding variables are level of education and socioeconomic status.

- Because the participants are related in some manner in a related-samples study, the responses in one sample are expected to be associated with the responses in the other sample. This relationship between the responses needs to be accounted for when finding t. This is done by including the ***covariance*** between the groups in the calculation of the $\sigma_M - \sigma_M$. Covariance is a measure of the extent to which two sets of scores vary together. Its formula is $2r(\sigma_1)(\sigma_2)$. The formula for $\sigma_M - \sigma_M$ in a related-samples t test is:

$$\sigma_M - \sigma_M = \sqrt{\sigma_{M1}^2 + \sigma_{M2}^2 - 2r\sigma_1\sigma_2}$$

- The formula for a related-samples t test then becomes:

$$t_{2-samp} = \frac{M_1 - M_2}{\sqrt{\sigma_{M1}^2 + \sigma_{M2}^2 - 2r\sigma_1\sigma_2}}$$

- Another method to calculate a related-samples t test uses the **direct-difference formula** (also called the **computational formula**). This formula allows you to do a related-samples t test without having to find r. In this formula, D represents the difference between all of the paired scores:

$$t_{2-samp} = \frac{\overline{D} - \mu_D}{\sqrt{\dfrac{\sum D^2 - \dfrac{(\sum D)^2}{n}}{n(n-1)}}}$$

- The degrees of freedom for a related-samples t test is calculated by subtracting 1 from the number of *pairs* in the samples: $df = n - 1$. Again, in this formula, n is the number of pairs, not the total number of participants.

Computational Exercises

1. Using the following data, calculate the standard error for the following repeated-measures design.

Pre	Post
39	38
56	13
28	29
47	12
16	45
56	38
52	16
12	42

2. You are hired by a dental company to evaluate the impact of a new teeth whitening process. The company informs you that it asked a sample of individuals to rate the whiteness of their teeth on a scale from 0 to 100. Then, the participants underwent the whitening process and were asked to then rate the whiteness of their teeth again. The company would like you to determine if the ratings significantly *improved*. Here are the data:

Before	After
45	57
82	94
67	76
21	64
51	50
32	48
37	47

   a. What are the independent and dependent variables for this study?

   b. State the null hypothesis and the research hypothesis.

   c. Can you reject the null hypothesis at $\alpha = .05$? Report your findings in APA format.

3. In a recent study of newlyweds, $n = 4$ married couples were asked to rate their marital satisfaction (-10 to 10 scale) after two years of marriage. The purpose of the study was to determine if couples differ on the perception of their marriage, so husbands and wives were asked independently, but the responses were matched across the couples. Here are the data:

Husband	Wife
10	9
8	7
2	4
4	6

   a. What are the independent and dependent variables for this study?

   b. State the null hypothesis and the research hypothesis.

   c. Can you reject the null hypothesis at $\alpha = .01$? Report your findings in APA format.

4. As the new head of a clothing company, you are asked to determine if women have a preference for long skirts or pants. You collect a sample of $n = 4$ women and ask them to rate their preference for long skirts and pants on a 1-5 scale. Here are the data:

Long Skirts	Pants
5	2
3	5
1	3
4	5
2	1
3	4

   a. What type of t test should be used?

   b. State the null hypothesis and the research hypothesis.

   c. Can you reject the null hypothesis at $\alpha = .05$? Report your findings in APA format.

5. A nutritionist is trying to determine if a high carbohydrate diet is more or less beneficial than a high protein diet. He recruits a sample of $n = 9$ individuals to go on a carbohydrate diet and another $n = 9$ to go on a protein diet. In order to reduce the influence of confounding variables, the nutritionist matches people across the samples according to health prior to starting to diet. He then measures the amount of weight loss over the next two weeks. Here are the results:

Carbohydrate	Protein
2	7
1	5
3	5
2	4
1	5
2	4
2	8
3	7
1	9

   a. What type of t test should be used?

   b. State the null hypothesis and the research hypothesis.

c. Can you reject the null hypothesis at $\alpha = .01$? Report your findings in APA format.

Computational Answers

1. $\sqrt{\dfrac{6437 - \dfrac{5329}{8}}{8(7)}} = 103.05.$

2a. Independent Variable: Teeth Whitening (Pre/Post)

Dependent Variable: Rating of whiteness of teeth

2b. Null: Self-ratings of teeth whiteness do not improve after the whitening process

Alternative: Self-ratings of teeth whiteness do improve after the whitening process

2c. Mean Difference = 14.43; Standard Error = 26.61

$$t_{2-samp} = \frac{14.43 - 0}{\sqrt{\dfrac{2575 - \dfrac{10201}{7}}{7(6)}}} = 0.54$$

Retain the null. Ratings of teeth whiteness did not significantly improve after using the new whitening process, $t(6) = 0.54$, $p > .05$.

3a. The independent variable is couple (husband/wife dyad). The dependent variable is marital satisfaction.

3b. Null: There is no difference between husband's and wife's marital satisfaction

Alternative: There is a difference between husband's and wife's martial satisfaction

3c. Mean Difference = –0.5; Standard Error = 0.75

$$t_{2-samp} = \frac{-.5 - 0}{\sqrt{\dfrac{10 - \dfrac{4}{4}}{4(3)}}} = -0.58$$

Retain the null. Martial satisfaction is not significantly different between husbands and wives, $t(3) = -0.58$, $p > .05$.

4a. Use a related-samples t test.

4b. Null: Women do not have a preference for skirts or pants

Alternative: Women do have a preference for skirts or pants

4c. Mean Difference $= -0.33$; Standard Error $= 0.80$

$$t_{2-samp} = \frac{-.33 - 0}{\sqrt{\frac{20 - \frac{4}{6}}{6(5)}}} = -0.41$$

Retain the null. There is no preference for skirts or pants, $t(5) = -0.41$, $p > .05$.

5a. Use a related-samples t test.

5b. Null: High carbohydrate and high protein diets will equally reduce weight

Alternative: High carbohydrate and high protein diets will reduce significantly different amounts of weight.

5c. Mean Difference $= -4.11$; Standard Error $= 0.46$

$$t_{2-samp} = \frac{-3.57 - 0}{\sqrt{\frac{185 - \frac{1369}{9}}{9(8)}}} = -9$$

Reject the null. High protein diets are more effective than high carbohydrate diets at reducing weight, $t(8) = -9.00$, $p < .05$.

True/False Questions

1. Covariance is a measure of the extent that two scores vary together.

2. In a repeated-measures study in the social sciences, it is impossible for the sample sizes in the groups to be unequal.

3. Participants were divided into two samples such that they were matched on the severity of their depression. These samples are considered independent because they contain different people.

4. The standard error in a related-samples t test must be adjusted to account for the relationship between the participants in the samples.

5. A repeated measures t test uses difference scores rather than raw scores.

True/False Answers

1. True
2. False
3. False
4. True
5. True

Short-Answer Questions

1. You are in charge of determining if a new TV show that is targeted at brothers and sisters is enjoyed by both siblings. As a result, you screen the show with $n = 20$ boys and their respective $n = 20$ sisters. In conducting your analysis, should you treat these samples as independent? Why or why not?

2. What is one of the major advantages of using a related-samples t test?

3. How are the formulas for degrees freedom different for independent-sample t tests and related-samples t tests? What is the cause of this difference?

4. What is the impact on the overall denominator of a repeated-measures t test of including a covariance?

5. Samantha's elementary school is having a candy-selling competition among the different classes in her grade. Samantha's aunt is a statistician and wants to determine if any of the classes sell significantly more candy. She decides to conduct a related-samples t test because all of the classes are in the same school. Is this correct? Why or why not?

6. The formulas for the three different types of two-sample t tests (equal sample sizes, unequal sample sizes, and related samples) all look strikingly different. However, appearances can be deceiving. Are the formulas for the different types of t tests really all that different? Why or why not?

Answers

1. No, these are not independent. Each participant is matched with a specific member in the other sample.

2. The influence of confounds is reduced because the same or similar participants are used in both groups.

3. The formula for an independent-sample test is $N - 2$, whereas the formula for a related-samples test is $n - 1$. This is because related samples deal with pairs of participants (or the same participant multiple times).

4. It decreases the size of the denominator, requiring a smaller difference in the numerator between the two groups to reject the null.

5. This is incorrect. Even though all of the students are in the same school, each class contains different students, and the students across the different classes were not matched in pairs.

6. The formulas for all the different two-sample t tests are all variations on the same formula. The weighting process used in calculating the standard error for a t test for unequal sample sizes could be applied to samples with equal sample sizes, but it would produce the same

results as if they were not weighted, and so it may be removed. The addition of a covariance term in a related-samples t test could be applied to the other t tests, except that it is expected to be 0. Therefore, we can remove it.

Multiple-Choice Questions

1. A sample of $n = 10$ has a mean difference of 6.7 and a standard deviation of the mean difference $= 3.8$. What is the statistical decision you would make using a nondirectional test?

   a. Reject the null at $\alpha = .05$, retain at $\alpha = .01$

   b. Retain the null at $\alpha = .05$ and at $\alpha = .01$

   c. Reject the null at $\alpha = .05$ and at $\alpha = .01$

   d. Retain the null at $\alpha = .05$, reject at $\alpha = .01$

2. For which of the following situations would a repeated-measures study be appropriate?

   a. Compare hours of sleep had by adolescents versus senior citizens

   b. Compare weight loss for individuals on a new diet to those on a different diet

   c. Compare salary levels for college graduates and those who did not graduate from college

   d. Compare reaction times before and after taking a pain medication

3. A team of researchers is interested in monitoring the influence of a new therapy for children with ADHD. They hope to see a decline in symptoms after two weeks of being enrolled in the therapy. Which of the following hypothesis tests would be best suited to determine if the therapy is effective?

   a. Normal deviate test

   b. One-sample t test

   c. Independent-sample t test

   d. Related-samples t test

4. For the study described in question 2, the researchers observe $n = 12$ children and obtain a mean difference of 4.5 in ADHD symptoms over the two-week period, with a standard deviation of the mean difference = 1.5. What is the statistical decision you would make using a nondirectional test?

   a. Reject the null at $\alpha = .05$, retain at $\alpha = .01$

   b. Retain the null at $\alpha = .05$ and at $\alpha = .01$

   c. Reject the null at $\alpha = .05$ and at $\alpha = .01$

   d. Retain the null at $\alpha = .05$, reject at $\alpha = .01$

5. For which of the following situations would a repeated-measures study be appropriate?

   a. Comparing marital satisfaction across partners in long-term committed relationships

   b. Comparing sales across two separate divisions of the same corporation

   c. Comparing real estate values on two blocks within the same neighborhoods

   d. Comparing sales of three separate pizza locations in the same city

Multiple-Choice Answers

1. B	4. A
2. D	5. A
3. D	

Module Quiz

1. A researcher is interested in assessing the effectiveness of a new medication for treating diabetes. She recruits a large sample of diabetics and randomly assigns them to two groups. The first group receives the medication, whereas the second group receives a placebo. Should the researcher should use a repeated-measures t test because all of the people in her sample are diabetics? Why or why not?

2. A company is trying to develop its next line of party favors. A focus group of 10 people is asked to rate their willingness to purchase a new type of electronic noisemaker. The focus group is then allowed to use the noisemakers and again asked to rate their willingness to purchase this favor. Should these researchers use a repeated-measures test? Why or why not?

3. The company from question 2 obtained the following data from the focus group. Ratings are on a 1-7 scale, with higher ratings indicating a greater desire to purchase the noisemaker. Based on these findings, does using the noisemaker increase the desire to purchase the favor? ($\alpha = .05$)

Before	After
3	2
3	7
6	7
2	5
5	7
4	7
1	5
1	2
3	4
7	5

4. Ms. Fiona is trying to improve the taste of her chicken pot pie. She obtained ratings on the international tastiness scale from $n = 8$ people who regularly attend her restaurant and had eaten her pot pie before she changed the recipe. She asks the same people to rate the tastiness of the pot pie after she changes the recipe. Based on the data below, does it seem like Ms. Fiona improved or ruined her famous recipe?

Old Recipe	New Recipe
26	78

14	79
22	88
45	77
39	98
29	76
16	84
20	89

5. A pharmaceutical company is trying to develop a medication to reduce cravings to drink alcohol in people who meet criteria for alcohol abuse. It obtains a sample of 7 individuals diagnosed with alcohol abuse and measures the number of drinks they have before and after taking the medication. Based on these findings, does the medication seem to affect the amount of drinks a person has?

Before	After
8	6
5	2
10	7
6	1
2	2
1	2
10	5

Quiz Answers

1. She should not use a repeated-measures *t* test because the participants in her sample are not the same individuals and they have not been matched on some criteria. Just because they are all part of the group of diabetics does not mean that she can use a repeated-measures test.

2. Yes, the researchers should use a repeated-measures test because the participants in the sample are the same individuals.

210

3. Mean difference = –1.6, Standard Error = 0.40

   Yes, it appears that using the noisemakers increases the desire to purchase them, $t(8) = -3.96$, $p < .05$.

   $$t_{2-samp} = \frac{-1.6 - 0}{\sqrt{\dfrac{62 - \dfrac{256}{10}}{10(9)}}} = -3.96$$

4. Mean difference = –57.25, Standard Error = 20.78

   Yes, it appears that the new recipe significantly improved the pot pie, $t(54) = -2.76$, $p < .05$.

   $$t_{2-samp} = \frac{-57.25 - 0}{\sqrt{\dfrac{27384 - \dfrac{209764}{8}}{8(7)}}} = -2.76$$

5. Mean difference = 2.43, Standard Error = 0.76

   Yes, it appears that the medication reduces the number of drinks that a person with alcohol abuse consumes, $t(5) = 3.22$, $p < .05$.

   $$t_{2-samp} = \frac{2.43 - 0}{\sqrt{\dfrac{73 - \dfrac{289}{7}}{7(6)}}} = 3.22$$

Module 23

Learning Objectives

- Distinguish between tabled and incurred alpha

- Understand the relationship between error and confidence

- Estimate parameters: point and interval

Module Summary

- The probability that we have not made a Type 1 error is called *confidence*. Confidence is calculated as $1 - \alpha$. Recall that α is the chance that we have made a Type 1 error. The precise level of α that is reported in a study is usually referred to as p. Actual confidence is calculated as $1 - p$.

- The social sciences will usually reject the null hypothesis when the probability of making a Type 1 error (p) is less than .05. However, this rule is not set in stone. When reporting your results, you are encouraged to report the actual p value rather than stating that your p value was below or above .05. This allows readers to make their own decisions about the difference between the two means rather than making it a dichotomous (significantly different/not significantly different) decision.

- After rejecting the null hypothesis, you can obtain an estimate of the range of the difference between the population means. This estimate is called the *parameter estimation*. There are two types of parameter estimates. The first is called a *point estimate*, in which the estimation of the difference between the two mains is a single number. For two-sample research, the point estimate is usually the actual difference between the sample means. The second is called an *interval estimate*, which estimates a range in which the true population difference most likely falls.

212

- The range established by the interval estimate is also referred to as a ***confidence interval***. Confidence intervals have an associated percentage that presents the confidence that the actual population mean difference falls within that range. For example, a 95% confidence interval of 5-10 means that you are 95% confident that the difference between two populations falls between the values of 5 and 10. The formula for calculating a confidence interval is:

$$CI = (M_1 - M_2) +/- (t_{\text{critical at } 1/2\,\alpha})(\sigma_{M-M})$$

Computational Exercises

1. Dr. Johnson wants to share the findings that students in his 1:00 P.M. class provide better evaluations than the students in his 8:30 A.M. class. To help him ensure that he can feel confident in his results, find the 95% confidence interval.

8:30 A.M.	1:00 P.M.
$n = 16$	$n = 24$
$M = 3.4$	$M = 5.7$
$SS = 89$	$SS = 95$

2. You are the head researcher for a TV station that has just completed conducting focus groups for a new show that is supposed to appeal to women. The results of an independent-sample t test support this finding, as men and women significantly differed on their interest in the show. The mean level of interest (on a scale of 1-10) for women was 7.8, whereas the mean level of interest for men was 3.4. The $\sigma_{M} - \sigma_{M}$ for the study was 1.5, with $N = 20$. In preparing to present this information to the owner of the TV station, you want to be able to report the possible population gender discrepancy in interest of this show.

    a. What would be the point estimate for the mean difference?

    b. What would be the 95% confidence interval for this difference?

    c. What would be the 99% confidence interval for this difference?

3. A bakery is trying to determine if it should sell its holiday bread all year round. It bakes a special batch of holiday bread in the summer and sells it for two weeks. It then compares sales of the bread during the two weeks of the holidays ($M = 21.20$) to the sales during the two weeks of the summer ($M = 23.12$). Using a $\sigma_M - \sigma_M = 2.3$ for the 14 days that the bread was sold across both the summer and holiday time periods ($n = 28$), what is the 95% confidence interval?

<u>Computational Answers</u>

1. CI $= (3.4 - 5.7) +/- (2.030)(0.71) = -5.44$ to 0.04

2a. 4.4

2b. CI $= (7.8 - 3.4) +/- (2.101)(1.50) = 7.55$ to 1.25

2c. CI $= (7.8 - 3.4) +/- (2.878)(1.50) = 8.72$ to 0.08

3. CI $= (21.20 - 23.12) +/- (2.101)(1.50) = -6.65$ to 2.81

True/False Questions

1. The application of confidence intervals for independent-sample tests is very similar to its application for repeated-measures t tests.

2. The formula used for to calculate a confidence interval differs for the type of a two-sample t test used (i.e. independent sample, repeated measures).

3. Confidence intervals for t tests display the expected range for the difference between the means.

<u>True/False Answers</u>

1. True   2. False   3. True

Short-Answer Questions

1. You have completed a study that has a $p = .03$. What is your confidence for this study? If this study was a comparison of two sample means, what is the probability that the difference between the means is attributed to sampling error? What is the probability that the difference is attributed to something other than sampling error? Which reason do you think is more likely for the difference between the means?

2. What is the problem with making a statistical decision a dichotomous decision?

3. How do confidence and a confidence interval differ?

Answers

1. Confidence would be .97. The probability that the difference between the means is attributed to sampling error is .03, or 3%. The probability that the difference between the means is attributed to something else (such as an independent variable) is .97, or 97%. There is a greater chance that the means differ because of another variable rather than sampling error.

2. It seems to make the answer appear as a black or white decision, that the IV did have an effect or the IV did not. We can never be this certain about any statistical decision. It is better to report a p value and allow each reader to make his or her own decision about the results.

3. Confidence refers to the probability that the difference between two samples is related to the independent variable. Confidence intervals refer to a range of values in which the difference between the two samples would be expected to fall if the study was done repeatedly.

Multiple-Choice Questions

1. Which of the following influence the size of a confidence interval for a t test?

   a. t_{crit} score

   b. The difference between the
   two sample means

   c. The standard error of the
   mean

   d. All of the above

2. The range of a confidence interval increases when the sample size _____ and the confidence level _____.

   a. Increases, Decreases

   b. Increases, Increases

   c. Decreases, Increases

   d. Decreases, Decreases

3. A new treatment for depression has been shown to be effective. On a commonly used depression inventory, the mean of the treatment group was 8 and the mean of the control group was 15, with a standard error of the mean difference of 3.2. If $N = 30$ participants were involved in this independent-samples t test, what is the 99% confidence interval for a nondirectional test?

   a. −1.83 to −15.83

   b. 1.83 to −15.83

   c. 10.2 to 4.7

   d. 11.2 to 4.8

Multiple-Choice Answers
   1. D
   2. C
   3. A

Module Quiz

1. A nutritionist compared a high protein diet to a high carbohydrate diet across $n = 10$ individuals. If the mean difference for the diet is 12 and the $\sigma_M - \sigma_M = 1.9$, what is the 95% confidence interval?

2. A nail salon wants to evaluate the 99% confidence interval for sales of manicures during a special and sales after a special. Across $N = 26$ days, the mean difference in sales between the promotional period and the nonpromotional period is 16 with an $\sigma_M - \sigma_M = 3.98$. What is the 99% confidence interval for the difference in sales?

3. A dance studio wants to determine the 95% confidence interval for which of two classes is more popular, Jazz or Latin. They compare attendance for two months for a total of $N = 10$ classes and find that the average difference in attendance between the two classes is 9.98 with an $\sigma_M - \sigma_M = 6.98$. What is the 99% confidence interval for the difference in attendance?

Quiz Answers

1. CI = 12 +/– (2.3060)(1.9) = 16.38 to 7.61

2. CI = 16 +/– (2.797)(3.98) = 27.13 to 4.87

3. CI = 9.98 +/– (3.355)(6.98) = 33.40 to –13.44

Module 24

Learning Objectives

- Know the assumptions underlying analysis of variance

- Distinguish between conditions under which it is appropriate to conduct a *t* test versus

  ANOVA

- Understand the relationship of within-group, between-group, and total variances

- Understand how the possible values of *F* follow from the distribution's shape

- Understand how changes in the within- or between-group variances influence the *F* statistic.

Module Summary

- In comparing three groups, you cannot use three separate *t* tests because that would unfairly

  increase the chances of rejecting the null. If you were to do a *t* test 100 times using $\alpha = .05$,

  you could theoretically expect to reject the null 5 times based on chance alone. Similarly, in

  doing three separate tests to compare three groups using $\alpha = .05$, the chances of falsely

  rejecting the null are raised.

- An *analysis of variance* or *ANOVA* is used to compare three or more groups while

  maintaining the level of Type 1 error. ANOVAs rely on the *F* statistic and search for

  differences in group variances rather than in group means. ANOVAs can be used to compare

  many types of groups, including living things (e.g., people, animals, plants) and inanimate

  objects (e.g., clothing, performances).

- Certain assumptions must be met in order to complete an ANOVA. The two primary

  assumptions are that the populations that samples are drawn from (1) are normally distributed

  and (2) have similar variances. These assumptions can be violated to a certain extent,

218

however, when the sizes of the samples are relatively large. An ANOVA is robust to these assumptions.

- ANOVA analyzes variances by splitting the variability of participants into two sections: (1) variability between groups, also known as between-group variability, and (2) variability within groups, also known as within-group variability. Between-group variability is the extent to which the treatment groups differ from each other. This variability is attributed to the independent variable, as it is hypothesized to be the main cause for group differences. Within-group variability is the extent to which members of the same group differ from each other. This variability is considered to be the result of random error, because members of the same group receive the same treatment.

- The primary analysis of an ANOVA entails splitting, or *partitioning*, the total variability into between-group and within-group variability and then comparing each piece to see if the between-group variance (treatment effect) significantly exceeds the within-group variance (error). In partitioning variance, you will need to calculate the means of each group as well as the overall mean (mean of all participants), which is referred to as the *grand mean*. Within-group deviations are the summed difference of each person from his or her own group mean ($X - M_g$). Between-group deviations are the summed difference of each person's group mean from the grand mean ($M_{tot} - M_g$). The total deviation is the summed difference of each person from the grand mean ($X - M_{tot}$).

- The sum of the squared deviation scores is referred to as the *sum of squares*. When any sum of squares is divided by the number of cases or scores, we have obtained a variance (or the average squared deviation). To find a variance for an ANOVA, you follow a very similar process to that of finding a variance for a sample. However, instead of dividing the sum of squares by n, as you would for a sample, you divide the variance by the degrees of freedom. Recall that the degrees of freedom correct for the bias that is inherent in using sample statistics to estimate population parameters.

- These variances of an ANOVA are called ***mean squares***. The mean square between groups (MS_{bet}) represents the variance primarily attributed to treatment differences between the groups, along with some random error. The mean square within groups (MS_{with}) represents the variance attributed solely to random error.

- The purpose of an ANOVA is to determine the amount of between-group variance relative to the amount of within-group variance. Thus, the formula for an ANOVA is:

$$F = \frac{MS_{bet}}{MS_{with}} = \frac{\text{Treatment} + \text{Error}}{\text{Error}} \approx \frac{\text{Treatment}}{\text{Error}}$$

- Similar to the t distribution, the F distribution is a family of distributions. Unlike the t distribution, it is positively skewed. An F ratio can never be negative because variances are always positive. Second, very few F ratios can fall below a score of 1. This would indicate that there is less variability between the groups than within the groups. The majority of situations in which there is no effect will have an equal amount of between- and within-group variability. Third, because an F ratio of 1 indicates no effect, it can be assumed that the majority of F ratios will cluster around a score of 1. Thus, $F = 1$ is the modal F ratio. As the treatment effect becomes more pronounced, the ratio will become progressively larger. Therefore, large F ratios occur less frequently and indicate a statistically significant treatment effect.

Computational Exercises

1. If the within-group deviation is 481.23 and the between-group deviation is 13,410.08, what is the total deviation?

2. If the total deviation for a data set is 5,698.65 and the between-group deviation is 2,573.36, what is the within-group deviation?

3. If the total deviation for a data set is 698.36 and the within-group deviation is 51.32, what is the between-group deviation?

4. If the $MS_{bet} = 586$ and the $MS_{with} = 987$, what is F?

5. If the $MS_{bet} = 87$ and the $MS_{with} = 24$, what is F?

Computational Answers

1. 13,891.31

2. 3,125.29

3. 647.04

4. 0.59

5. 3.62

True/False Questions

1. ANOVA uses a completely different method of inferential logic from that of a t test or normal deviate test.

2. ANOVA enables you to compare more than two groups while maintaining the same level of Type 1 error.

3. ANOVAs directly compare group mean differences.

4. ANOVA is robust to the assumption of normality with larger sample sizes.

5. When variance is partitioned in an ANOVA, it is divided into two equal parts.

6. The grand mean is the mean of all scores in the analysis.

7. Between-group variability contains only variability attributed to the treatment.

8. An F less than 1 has a very high chance of being significant.

True/False Answers

1. False	4. True	7. False
2. True	5. False	8. False
3. False	6. True	

Short-Answer Questions

1. What are the assumptions that must be met in order to use an ANOVA?

2. What does it mean that ANOVA is robust to its assumptions?

3. What is within-group variability attributed to? How does this differ from between-group variability?

4. Mean differences are compared when using a t test. What does an ANOVA compare?

5. Demonstrate how a single person's score can be partitioned into between-group and within-group variance.

6. Can an F ratio test be directional? Why or why not?

7. What is a mean square? Is it the same or different from a variance?

Answers

1. The first assumption is that the populations from which the samples were drawn are normally distributed. The second is that the populations from which the samples were drawn have equal variances.

2. It means that with large enough samples, the assumptions that underlie the use of ANOVA can be violated and the results will still be valid.

3. Within-treatment variability is attributed solely to random error. Between-treatment variability is attributed primarily to the independent variable (treatment groups). However, between-group variability also contains a small amount of random error.

4. An ANOVA compares variances. The variability between treatment groups is compared to the variability within each treatment group.

5. Total variability is the extent that the person's score differs from the grand mean (or the mean of all scores). Between-group variability is the extent that the mean of group the person belongs to differs from the grand mean. Finally, within-group variability is the extent to which the person's score differs from the mean of his or her group.

6. Although the question may be directional (does something increase or decrease?), the actual test itself is always positive. This is because it is based on squared values, which are always positive. Also, the ratio of between-group to within-group positive variances produces a highly skewed distribution. Thus, significant F ratios are always larger than 1.

7. A mean square is the average of the squared deviations. It is essentially the same as a variance; however, it is important to note that a mean square is an estimate of the population variance.

Multiple-Choice Questions

1. What is the F ratio of a treatment that had no effect?

    a. 20

    b. 15

    c. 78

    d. 1 or less

2. Which of these values can an F ratio never take?

    a. 1

    b. 1000

    c. 0.54

    d. –2

3. What other statistical term is the mean square most similar to?

223

a. Mean

    c. Variance

b. Standard Deviation

    d. Sum of Squares

4. In an ANOVA, a desirable (significant) F ratio contains a lot of _____ variability as compared to _____ variability

    a. Between-group, Within-group

    b. Between-subject, Within-subject

    c. Within-group, Between-group

    d. Within-subject, Between-subject

5. Between-group variability is a measure of

    a. the deviation of each score from the grand mean

    b. the deviation of each score from its group mean

    c. the deviation of each group mean from the grand mean

    d. variability attributed to random error

6. What information does an ANOVA provide?

    a. Whether or not there is a significant difference between groups at a set alpha level

    b. Which group had the significantly highest mean

    c. Which group had the significantly lowest mean

    d. The amount that two group means must differ in order to be considered significantly different

Multiple-Choice Answers

1. D

    4. A

2. D

    5. C

3. C

    6. A

Module Quiz

1. What other mathematical concept is most similar to the mean square?

2. In conducting a research study, what should the researcher attempt to do in recruiting the samples in an attempt to increase the F ratio?

3. What is the primary source of within-group variability?

4. Professor Jones wants to compare the overall GPA of his four classes. Why can't he use multiple t tests?

5. If the $MS_{bet} = 456$ and the $MS_{with} = 74$, what is F?

Quiz Answers

1. The mean square is a variance, as it is the average of the squared deviation scores.

2. The researcher can do a number of things. However, in the context of this chapter, the best thing a researcher could do is try to ensure that all of the participants in the sample are very similar. This will minimize the effect of any differences among the individuals and reduce the within-group variability.

3. Within-group variability represents the amount of error variance in the sample, largely view as random error variance.

4. He would be unfairly increasing the chances of obtaining a significant result. In other words, he would be inflating the chances of making a Type 1 error.

5. 6.16.

Module 25

Learning Objectives

- Understand the similar logic underlying various test statistics

- Determine the degrees of freedom

- Calculate a one-way ANOVA

- Use a table to interpret calculated F

- Display the results in an ANOVA summary table

- Report results in APA format

Module Summary

- Just as there were different types of t tests, there are different types of ANOVAs. The first ANOVA is a one-way ANOVA. *One-way ANOVAs* are used when there is only one independent variable.

- Much like all other hypothesis tests, a one-way ANOVA is used to answer the question of "is the difference among these groups a lot or a little?" As a result, it follows the prototypical formula of all other tests. However, the way in which you compare the differences between your obtained and expected values to the standardized random error is not as apparent. The obtained difference between the groups in the numerator is in terms of variance, rather than means. The expected difference between the groups is 0 (as it was with the t tests). The standardized random error in the denominator in ANOVA refers to the within-group variance rather than the standard deviation. It is still the variability that is due to random chance.

- The first part of calculating a one-way ANOVA is partitioning the total sum of squares (SS) into its between-group and within-group portions. First, the total SS should be found with the following formulas:

226

Deviation Method:

$$SS_{tot} = \sum_1^N (X - M_{tot})^2$$

Raw Score Method:

$$SS_{tot} = \sum_1^N X^2 - \sum_1^k \frac{(\sum [X_{tot}])^2}{N}$$

Next, the between-group SS is found with the following formulas: (k = number of groups):

Deviation Method:

$$SS_{bet} = \sum_1^k n_g (M_g - M_{tot})^2$$

Raw Score Method:

$$SS_{bet} = \sum_1^N \frac{(\sum [X_g])^2}{n_g} - \frac{(\sum [X_{tot}])^2}{N}$$

Finally, the within-group SS is found with the following formulas:

Deviation Method:

$$SS_{with} = \sum_1^k \sum_1^n (X - M_g)^2$$

Raw Score Method:

$$SS_{with} = \sum_1^N X^2 - \sum_1^k \frac{(\sum [X_g])^2}{n_g}$$

There is a great deal of math that is associated with partitioning the variance, so it is important to be organized when you are conducting your ANOVA. Also, there is a "safety net" in place that enables you to check your work: $SS_{tot} = SS_{bet} + SS_{with}$. You should always add your between-groups SS and within-groups SS after conducting an ANOVA to ensure that they equal the total SS.

- After the variance has been partitioned, the next step is to find the appropriate degrees of freedom. The formulas for the degrees of freedom are as follows:

$$df_{bet} = \text{\# of groups} - 1 = k - 1$$
$$df_{with} = \text{\# of subjects} - \text{\# of groups} = N - k$$
$$df_{tot} = \text{\# of subjects} - 1 = N - 1$$

- After obtaining the degrees of freedom and the SS, we can now find the mean square by dividing each SS by its corresponding degrees of freedom.

$$MS_{bet} = \frac{SS_{bet}}{df_{bet}}$$
$$MS_{with} = \frac{SS_{with}}{df_{with}}$$

- The final step in this process is to calculate the F value, which is done by dividing the MS_{bet} by the MS_{with}.

- Statistical significance for an F value is both similar to and different from that of a t test. It is similar in that it continues to compare deviations from an expected result to random error. However, it differs in that it cannot use the normal curve or t curve to determine if this difference is a lot or a little. As a result, a different table is used for F ratios, which can be found in Appendix D. Remember, a significant F ratio will be one that is very different from a value of 1. The table requires the use of the two degrees of freedom that were used to calculate the F ratio, the df_{bet} and the df_{with}. The obtained F ratio is then compared to the critical tabled F ratio. If the F ratio from the ANOVA (obtained F) is larger than the one from the table (critical F), the null is rejected. Otherwise, the null is retained.

- The results of an ANOVA are presented in APA format in the following fashion: $F(df_{bet}, df_{with}) = F$ obtained, $p <$ or $>$ error value.

- All of the calculations and steps involved in an ANOVA can be difficult to organize and present in a neat format. An efficient organization of all the pertinent information is found in an *ANOVA summary table,* which appears as follows:

Source	SS	df	MS	F
Between-groups	SS_{bet}	df_{bet}	SS_{bet}/df_{bet}	MS_{bet}/MS_{with}
Within-groups	SS_{with}	df_{with}	SS_{with}/df_{with}	
Total	SS_{tot}	df_{tot}		

Computational Exercises

1. Partition the variability for the following scores; that is, find the SS_{bet}, SS_{with}, and SS_{tot}.

Group 1	Group 2	Group 3
9	5	8
1	10	6
5	3	9
3	8	3
7	2	1

2. Find the MS values for the data in question 1. Without finding the F ratio, do you suspect it will be a significant value? Why or why not?

3. A music research group is interested in determining which of the three most popular music acts has the most regular concert attendance. The researchers interview 6 fans of each music act and ask them how many concerts they have attended. Here are the data they obtain:

Act 1	Act 2	Act 3
4	4	0
3	1	1
5	2	0
2	1	1

3	4	0
5	3	2

a. What are the null and alternative hypotheses for this study?

b. Make a statistical decision at the .01 level to determine if there is a difference among the acts.

c. Write the results in APA format.

4. Three different statistics professors are asked by their department to determine if they are teaching reliably (this would be determined by there being no difference in the performance of the students across the classes). The statistics professors decide to take a random sample of grades from each of their classes and conduct an ANOVA to determine if there are any differences among the classes. Here are the data they obtain:

Class 1	Class 2	Class 3
85	92	76
93	80	83
85	93	77
98	75	94
80	92	100

a. What are the null and alternative hypotheses for this study?

b. Make a statistical decision at the .01 level to determine if there is a significant difference among the teachers.

5. Three nail salons are competing to determine which salon can provide the longest lasting manicure. They compare the length of time it takes for $n = 5$ customers to require a new manicure. Here are the data they obtain:

Salon 1	Salon 2	Salon 3
27	23	17
24	27	21

20	21	24
26	12	25
15	10	16

a. What are the null and alternative hypotheses for this study?

b. Make a statistical decision at the .05 level determining if there is a significant difference among the salons and write the results in APA format.

6. A construction company is using a new adhesive in its new building. The company wants to determine on which surface the adhesive works best: wood, steel, or concrete. To determine this, it uses the adhesive on each surface four different times and determines how much pressure (in pounds) can be applied before the adhesive fails. Here are the data:

Wood	Steel	Concrete
97	141	94
88	132	101
98	134	103
84	129	91

a. What are the null and alternative hypotheses for this study?

b. Make a statistical decision at the .05 level determining if there is a significant difference among the adhesives.

c. Write the results in APA format.

7. Researchers at a pharmaceutical company are testing a new drug for alcoholism. They obtain a sample of $N = 21$ alcoholics and randomly assign them to one of three groups: treatment, control, and placebo. Participants are asked to take the medication for 4 weeks, after which the effectiveness of their liver enzymes will be assessed on a 0-100 scale. Here are the data they obtain:

Drug	Control	Placebo
92	26	15
93	33	53
89	53	55

95	45	22
36	45	84
79	71	42
89	24	90

    a. What are the null and alternative hypotheses for this study?

    b. Make a statistical decision at the .05 level determining if there is a significant difference among the conditions.

    c. Write the results in APA format.

8. Using the study from question 7, create an ANOVA source table.

9. A cognitive psychologist is interested in studying the influence of stress on reaction time. She obtains a sample of $N = 30$ participants and randomly assigns them to one of three groups. The groups are asked to watch a short film that is neutral (calm), induces some stress (stressed), or induces a lot of stress (highly stressed). The researcher obtains the following information from her participants:

	Calm	Stressed	Highly Stressed
Mean	230	450	497
SS	5478	5347	5317

    a. The $SS_{tot} = 18,542$. Create a source table that will address the primary hypothesis of the researcher. What are the null and alternative hypotheses for this study?

    b. Make a statistical decision at the .05 level.

    c. Write the results in APA format.

10. *Pet Lovers* magazine is conducting its annual poll to determine how much affection different pet owners have for their pets. It sends out a survey asking pet owners to rate their love for their animals on a 0-50 scale and have 5 different categories of pet: Dog, Cat, Fish, Turtle, and Rabbit. A grand total of 100 responses come in, with an equal number of responders falling in each pet category. Here are the data obtained from the survey:

	Dog	Cat	Fish	Turtle	Rabbit
Mean	39	34	30	24	28
SS	210	324	245	207	547

The SS_{tot} = 2000. Create a source table that will address whether there are differences between the levels of love different pet owners have for their pets.

    a. What are the null and alternative hypotheses for this study?

    b. Make a statistical decision at the .05 level.

    c. Write the results in APA format.

Computational Answers

1. $SS_{bet,}$ = 0.93; SS_{with} = 130.4; SS_{tot} = 131.33.

2. MS_{bet} = 0.47; MS_{with} = 10.87. You can expect there to not be a significant F ratio because the MS_{bet} is very small compared to the MS_{with}.

3a. Null: There is no difference in the number of concerts attended by the fans of each music act.

Alternative: There is a difference in the number of concerts attended by the fans of each music act.

3b. *Means*: 1 = 3.67; 2 = 2.5; 3 = 0.67

SS: SS_{bet} = 27.44; SS_{with} = 20.17; SS_{tot} = 47.61

df: df_{bet} = 2; df_{with} = 15; df_{tot} = 17

MS: MS_{bet} = 13.72; MS_{with} = 1.34

F = 10.27

Reject the null.

3c. There was a significant difference among the amount of concerts attended across the three music acts, $F(2, 15) = 10.27, p < .01$.

4a. Null: There is no difference in grades of the students across the three classes.

Alternative: There is a difference in grades of the students across the three classes.

4b. *Means*: 1 = 88.2; 2 = 86.4; 3 = 86

   SS: SS_{bet} = 13.73; SS_{with} = 934; SS_{tot} = 947.73

   df: df_{bet} = 2; df_{with} = 12; df_{tot} = 14

   MS: MS_{bet} = 6.87; MS_{with} = 77.83

   F = 0.09

   Retain the null.

4c. There is no significant difference among the performance of students in the three different

   classes, $F(2, 14)$ = .09, p > .05.

5a. Null: There is no difference in the length in which a manicure lasts across the nail salons.

   Alternative: There is a difference in the length in which a manicure lasts across the nail

   salons.

5b. *Means*: 1 = 22.4; 2 = 18.6; 3 = 20.6

   SS: SS_{bet} = 36.13; SS_{with} = 375.6; SS_{tot} = 411.73

   df: df_{bet} = 2; df_{with} = 12; df_{tot} = 14

   MS: MS_{bet} = 18.07; MS_{with} = 31.3

   F = 0.58

   Retain the null.

   There is no significant difference in the length of time in which a manicure lasts from any of

   the salons. $F(2, 14)$ = 0.58, p > .05.

6a. Null: There is no difference in the strength of the adhesive on different surfaces.

   Alternative: There is a difference in the strength of the adhesive on different surfaces.

6b. *Means*: Wood = 91.75; Steel = 134; Concrete = 97.25

   SS: SS_{bet} = 4221.17; SS_{with} = 315.5; SS_{tot} = 4536.67

df: $df_{bet} = 2$; $df_{with} = 9$; $df_{tot} = 11$

MS: $MS_{bet} = 2110.58$; $MS_{with} = 35.06$

$F = 60.21$

Reject the null.

6c. There is a significant difference in the strength of the adhesive on different surfaces, $F(2, 9) = 60.21$, $p < .05$.

7a. Null: There is no difference in the effectiveness of liver enzymes among the groups.

Alternative: There is a difference in the effectiveness of liver enzymes among the groups.

7b. *Means*: Drug = 81.85; Control = 42.43; Placebo = 51.57

SS: $SS_{bet} = 5962.67$; $SS_{with} = 9098.29$; $SS_{tot} = 15060.95$

df: $df_{bet} = 2$; $df_{with} = 18$; $df_{tot} = 20$

MS: $MS_{bet} = 2981.33$; $MS_{with} = 505.46$

$F = 5.90$

Reject the null.

7c. There is a significant difference in the effectiveness of liver enzymes among the groups, $F(2, 18) = 5.90$, $p < .05$.

8.

	SS	df	MS	F
Between-groups	5,962.67	2	2981.33	5.9
Within-groups	9,098.29	18	505.46	
Total	15,060.95	20		

9a. Null: Stress level has no impact on reaction time.

Alternative: Stress level has an impact on reaction time.

9b.

	SS	df	MS	F
Between-groups	2400	2	1200	2.01
Within-groups	16142	27	597.85	
Total	18542	29		

Retain the null.

9c. Stress level does not significantly impact reaction time, $F(2, 27) = 2.01, p > .05$.

10a. Null: The type of pet does not affect the amount of love expressed by the owner.

Alternative: The type of pet does affect the amount of love expressed by the owner.

10b.

	SS	df	MS	F
Between-groups	467	4	116.75	7.23
Within-groups	1533	95	16.14	
Total	2000	99		

Reject the null.

10c. The amount of love a pet owner expresses toward a pet is related to the type of pet,

$F(4, 95) = 7.23, p < .05$.

True/False Questions

1. A one-way ANOVA compares differences within one group.

2. The F distribution appears very similar to the Z and t distributions.

3. In a perfect situation, all individuals in the same group would have the exact same score.

4. Three groups have different group means, but there is a great amount of variability in each group. This situation will definitely yield a significant F ratio.

5. A strong treatment effect will be related to a large numerator in computing the F ratio.

True/False Answers

1. False

2. False

3. True

4. False

5. True

Short-Answer Questions

1. Why do you divide the *SS* by the *df* in an ANOVA, instead of by *n*?

2. Explain why the *F* distribution appears as positively skewed.

3. What would cause an *F* ratio to be statistically significant?

4. The numerator in an *F* ratio contains random error. Where did this random error come from?

5. In writing the results of an ANOVA in APA format, which degrees of freedom values must be reported?

6. What are the primary steps in conducting an ANOVA?

Answers

1. The *SS* is divided by the *df* because the variances in an ANOVA are estimated population variances and sample statistics tend to underestimate population parameters. Because the *df* subtracts at least one from the denominator, it enables the resulting mean square to be a more accurate population estimate.

2. The *F* distribution is positively skewed because the majority of *F* ratios (those that do not indicate a significant difference) will be close or equal to 1. It is very unusual for an *F* ratio to be less than 1, so very few values will fall below 1. However, as the differences among the groups increase, the *F* ratio becomes increasingly larger (and the probability of obtaining such a large *F* ratio decreases).

3. A significant *F* ratio is the result of a large amount of between-group variability (large group differences) and a minimal amount of within-group variability (very few differences among those in the same groups).

4. This random error comes from additional confounds that unfortunately cannot be addressed. It is merely a fact of life that this measurement will contain random error. This is why we can never be certain about the conclusion we draw from our hypothesis test.

5. The degrees of freedom for the numerator (between) and those for the denominator (within).

6. The first step is to partition the variability (sum of squares). The second step is to find the degrees of freedom for each section of variability. Third, the mean squares are found. Finally, the F ratio is calculated.

Multiple-Choice Questions

1. In a data set containing three groups, the total SS is 431 and the between-groups SS is 123. What is the within-groups SS?

   a. 308

   b. 554

   c. 234

   d. 123

2. For a data set containing three groups, $SS_{tot} = 8792$; $SS_{group\ 1} = 427$; $SS_{group\ 2} = 574$; $SS_{group\ 3} = 177$. What is the SS_{bet}?

   a. 1178

   b. 7614

   c. 5148

   d. 8365

3. A research study has a total of $N = 400$ participants across 8 groups. If there are equal numbers of participants in each group, what are the degrees of freedom for this study?

   a. $df_{bet} = 7$; $df_{with} = 400$; $df_{tot} = 399$

   b. $df_{bet} = 8$; $df_{with} = 392$; $df_{tot} = 399$

   c. $df_{bet} = 24$; $df_{with} = 247$; $df_{tot} = 399$

   d. $df_{bet} = 7$; $df_{with} = 392$; $df_{tot} = 399$

4. A research study containing three groups has an $n = 14$ participants per group. What are the degrees of freedom for this study?

a. $df_{bet} = 2$; $df_{with} = 39$; $df_{tot} = 42$

b. $df_{bet} = 2$; $df_{with} = 39$; $df_{tot} = 41$

c. $df_{bet} = 39$; $df_{with} = 2$; $df_{tot} = 42$

d. $df_{bet} = 42$; $df_{with} = 39$; $df_{tot} = 2$

5. Which value of the mean square between groups is most indicative of no group differences?

    a. 1   c. 0

    b. 7   d. 2

6. A research study has $N = 12$ participants in 3 groups (an equal number of participants per group). What is the F critical value at $\alpha = .01$?

    a. 3.88   c. 6.99

    b. 4.26   d. 8.02

7. The research study in question 12 yields an F value of 6.32. Which of the following correctly shows the result in APA format?

    a. $F(2, 12) = 6.32$, $p < .01$ c. $F(2, 9) = 6.32$, $p > .01$

    b. $F(3, 12) = 6.32$, $p > .01$ d. $F(2, 9) = 6.32$, $p < .01$

8. A study of the effectiveness of a new material for contact lenses has $n = 10$ people per group in four different groups. The results indicate that the $SS_{tot} = 2147$ and $SS_{bet} = 354$. What is the MS_{with}?

    a. 1793   c. 49.81

    b. 36.21   d. 88.5

9. Using the information from question 14, what is the F ratio?

    a. 88.5   c. 2.37

    b. 49.81   d. 5.23

10. Using the previous question's information, what conclusion would you draw with $\alpha = .05$?

    a. $F(4, 10) = 2.37, p > .05$ c. $F(3, 36) = 2.37, p > .05$

    b. $F(4, 10) = 5.23, p > .05$ d. $F(3, 36) = 5.23, p > .05$

11. Below is the source table for a study on depression comparing two different treatments to a placebo group. What is the F ratio?

	SS	df	MS	F
Between-groups	782			
Within-groups		33		
Total	2147			

    a. 9.45   c. 8.41

    b. 7.25   d. 6.23

12. Using the information from question 11, how many participants were involved in the study?

    a. 36   c. 39

    b. 34   d. 37

13. A study was done comparing how well students performed on a final exam in three different rooms: the original room in which they had the class, a room different from the one in which they had the class, or a different room but they were thinking about the original room. In all, 60 students were involved, with an equal number of students in each group. Here are the SS values for each group: Same = 475; Different = 427; Think = 412. Using the following source table, what is the F value of this study?

	SS	df	MS	F
Between-groups	1159	2	579.5	
Within-groups	1314	57	230.05	
Total	2473	59		

    a. 32.21   c. 230.05

    b. 14.21   d. 25.14

14. Which would the correct way of writing the obtained result from the previous question in APA format?

 a. $F(2, 59) = 25.14, p < .05$ c. $F(3, 59) = 230.05, p > .05$

 b. $F(2, 57) = 25.14, p < .01$ d. $F(2, 59) = 25.14, p > .05$

Multiple-Choice Answers

1. A   8. C

2. B   9. C

3. D   10. C

4. B   11. A

5. C   12. A

6. C   13. D

7. C   14. B

Module Quiz

1. Partition the variability for the following scores. (Find the *SS* between, within, and total.)

Group 1	Group 2	Group 3
3	10	8
2	8	10
6	7	8
5	3	10
1	7	9
2	2	9

2. You are conducting research on a new therapy for social anxiety by comparing those who receive the therapy to those who do not. However, you want to avoid the possibility that people are improving just because they are interacting with another person. To address this confound, you add a group of participants who will meet with a therapist on a schedule similar to that of the social anxiety group, but will not receive therapy. This group will be

241

referred to as the attention-control group. Here are the data you obtain from a measure of

social anxiety with a scale of 0-20:

Therapy	Control	Attention-Control
2	19	14
7	13	15
5	20	13
4	14	12
2	12	9

   a. What are the null and alternative hypotheses for this study?

   b. Make a statistical decision at the .05 level to determine if there is a significant

       difference among the different treatments.

   c. Write the results in APA format.

3. A local police department has divided the city into three sections. The police chief wants to

    determine if officers are biased in the number of parking tickets they write in each section.

    Here are the data for the number of parking tickets given out by the 6 officers who work in

    each section of the city:

Area 1	Area 2	Area 3
8	3	1
4	7	2
6	0	7
8	2	6
6	7	5
4	5	0

   a. What are the null and alternative hypotheses for this study?

   b. Make a statistical decision at the .05 level to determine if there is a significant

       among the different sections.

   c. Write the results in APA format.

4. A prominent ice cream company is trying to discover which ice cream flavor is most

    preferred by its customers. It has narrowed the choices to mint chocolate chip, butter pecan,

    pistachio, and cookie dough. It decides to investigate daily sales of each flavor over the course of two weeks ($n = 14$ days) to determine which flavor is most preferred. Here are the data the company obtained:

	Mint Chocolate Chip	Butter Pecan	Pistachio	Cookie Dough
Mean	23	24	31	37
SS	78	84	81	92

   a. What are the null and alternative hypotheses for this study?

   b. The $SS_{tot} = 378$. Create a source table that will address if there is a significant different among the preferences for the flavors. Make a statistical decision at the .05 level.

   c. Write the results in APA format.

5. A school chancellor has very limited funding for the schools in her district. She is able to narrow down her choices to the three schools with the most need but is unable to determine which school would benefit from the most funding. To help with her decision, she consults the annual reading scores of each school over the past 7 years to determine which school would benefit the most. Here are the data she obtains:

	School 1	School 2	School 3
Mean	63	61	58
SS	201	213	278

   a. What are the null and alternative hypotheses for this study?

   b. The $SS_{tot} = 945$. Create a source table that will help the chancellor decide which school will receive the funding. Make a statistical decision at the 0.05 level.

   c. Does it appear that there is one school with greater need? Write the results in APA format.

Quiz Answers

1. $SS_{bet} = 102.11$; $SS_{with} = 69.67$; $SS_{tot} = 171.78$.

2a. Null: There is no difference in social anxiety among the groups.

Alternative: There is a difference in social anxiety among the groups.

2b. *Means*: Treatment = 4; Control= 15; Attention-Control = 12.6

SS: SS_{bet} = 362.53; SS_{with} = 92.4; SS_{tot} = 454.93

df: df_{bet} = 2; df_{with} = 12; df_{tot} = 14

MS: MS_{bet} = 181.27; MS_{with} = 7.7

$F = 23.54$

Reject the null.

2c. There was a significant difference among the different groups, $F(2, 12) = 23.54, p < .05$.

3a. Null: There is no difference in the number of tickets distributed by police in the three areas.

Alternative: There is a difference in the number of tickets distributed by police in the three areas.

3b. *Means*: 1 = 6; 2 = 4; 3 = 3.5

SS: SS_{bet} = 21; SS_{with} = 97.5; SS_{tot} = 118.5

df: df_{bet} = 2; df_{with} = 15; df_{tot} = 17

MS: MS_{bet} = 10.5; MS_{with} = 6.5

$F = 1.62$

Retain the null.

3c. There was no significant difference among the tickets distributed by police across the three areas of the city, $F(2, 15) = 1.62, p > .05$.

4a. Null: There is no difference among the preferences for the different flavors of ice cream.

Alternative: There is a difference among the preferences for different flavors of ice cream.

4b.

	SS	df	MS	F
Between-groups	43	3	14.33	0.43
Within-groups	335	10	33.50	
Total	378	13		

Retain the null.

4c. There is no difference among the preferences for the different flavors of ice cream, $F(3, 10) =$ 0.43, $p > .05$.

5a. Null: There is no difference in the scores on the reading test among the schools.

Alternative: There is a difference in the scores on the reading test among the schools.

5b.

	SS	df	MS	F
Between-groups	253	2	126.5	0.73
Within-groups	692	4	173	
Total	945	6		

Retain the null.

5c. There is no difference in need among the three schools based on the reading test, $F(2, 4) =$ 0.73, $p > .05$.

Module 26

Learning Objectives

- Distinguish between the information provided by an *F* test and a Tukey HSD

- Know when it is appropriate to calculate a Tukey HSD

- Calculate a Tukey HSD

- Construct a matrix of group mean differences

- Report results in APA format

Module Summary

- An ANOVA (*F* test) tells you that there is a significant difference among the groups that were compared; however, it does not specify which groups are significantly different. To determine which groups are significantly different from one another, you must conduct a *post hoc* test. If you do not have a significant *F* ratio, then a post hoc test is unnecessary.

- There are many different types of post hoc tests, but all have the same function: determining significant differences among the groups that were compared in an ANOVA. They are similar to conducting a series of *t* tests in which you compare pairs of groups. Given this, they are often referred to as *pairwise* comparisons. Unlike multiple *t* tests, post hoc tests do not increase the chance of making a Type 1 error. This is because post hoc tests use a fixed error rate. In other words, it "spreads the α around" the multiple comparisons to ensure that the probability of making a Type 1 error remains stable.

- One post hoc test is called the Tukey HSD, or *Honestly Significant Difference*. If the difference between two group means exceeds the value obtained by the HSD, then they are considered significantly different. If the difference between two group means is less than the value of the HSD, then they are not significantly different. For example, if the HSD = 3, then

246

means of 6 and 4 would not be significantly different, because the difference between these means is 6 – 4 = 2, which is less than the HSD. However, means of 6 and 2 would be significantly different because their difference of 6 – 2 = 4, which exceeds the value of the HSD. The formula for the HSD is as follows:

$$\text{HSD} = q\sqrt{\frac{MS_{with}}{n_g}}$$

- In the formula above, a new statistic is presented, the **q** statistic, which refers to the **studentized ranged statistic.** The value q is a correction factor that helps the post hoc test maintain a steady level of Type 1 error. To find q, refer to the table in Appendix E. The value of q is found by using k (the number of groups) and the df_{with}.

- It is convenient to display pairwise mean differences in a grid as follows. This allows you to quickly compare obtained differences to the HSD:

	Group 1	Group 2	Group 3
Group 1	0		
Group 2		0	
Group 3			0

Computational Exercises

1. Using the tables below, determine which groups are significantly different at $\alpha = .05$.

	SS	df	MS	F
Between	214	3	71.33	10.90
Within	314	48	6.54	
Total	528	51		

	1	2	3	4
Mean	32	24	67	54
n	13	13	13	13

2. Using the tables below, determine which groups are significantly different at α = .05.

	SS	df	MS	F
Between	327	2	163.50	9.66
Within	457	27	16.93	
Total	784	29		

	1	2	3
Mean	20	34	45
n	10	10	10

3. Here are the data from Computational Exercise 9 of the previous module. Determine where

the significant differences fall at α = .05.

	Calm	Stressed	Highly Stressed
Mean	230	450	497
SS	5478	5347	5317

	SS	df	MS	F
Between	2400	2	1200	2.01
Within	16142	27	597.85	
Total	18542	29		

4. Using the following tables, compute an HSD at α = .05. There were equal numbers of

participants per group.

	1	2	3	4
Mean	47	42	34	21

	SS	df	MS	F
Between	578	3	192.67	6.24
Within	987	32	30.84	
Total	1565	35		

5. Using the following tables, compute the Tukey HSD test at α = .05.

	1	2	3
Mean	47	42	34
n	12	12	12

248

	SS	df	MS	F
Between	478	2	239	5.43
Within	1453	33	44.03	
Total	1931	35		

Computational Answers

1. $3.51\sqrt{\dfrac{6.54}{13}} = 2.67$. All of the groups are significantly different from one another.

2. $3.51\sqrt{\dfrac{16.93}{10}} = 4.57$. All of the groups are significantly different from one another.

3. $3.51\sqrt{\dfrac{597.85}{10}} = 27.14$. All of the groups are significantly different from one another.

4. $3.84\sqrt{\dfrac{30.84}{9}} = 7.11$. Groups 1 and 2 are significantly different from Groups 3 and 4. Group

3 and Group 4 are significantly different from each other as well.

5. $3.48\sqrt{\dfrac{44.03}{12}} = 6.67$. Groups 1 and 2 are significantly different from Group 3.

True/False Questions

1. A post hoc test is exactly like doing multiple t tests.

2. Although there are many post hoc tests, they all provide the same results.

3. The HSD test can be used only when the groups have equal numbers of participants.

4. The q statistic is used to maintain the level of Type 1 error in a Tukey HSD.

5. The value of q is found by using the df_{bet} and the df_{with}.

6. Means with a difference less than that obtained by the HSD are significantly different.

7. In an HSD test, the denominator is the number of participants in the study.

True/False Answers

1. False   2. False   3. True

4. True	6. False
5. False	7. False

Short-Answer Questions

1. Why are post hoc tests necessary?

2. Chester has just completed an ANOVA, with the result of $F(2, 10) = 1.23$, $p > .05$. Should he do a post hoc test? Why or why not?

3. Why is an ANOVA with post hoc tests preferable to conducting numerous t tests?

4. What is the purpose of the q statistic?

5. What pieces of information are used to determine the q statistic?

6. What is an HSD?

<u>Answers</u>

1. They determine which groups are significantly different in an ANOVA. This information is not provided from the original F test.

2. He should not, because he does not have a significant F ratio.

3. ANOVA and post hoc tests maintain a constant level of Type 1 error, whereas multiple t tests will raise the probability of making a Type 1 error.

4. The q statistic is used in an HSD. It helps to ensure that the chances of making a Type 1 error are not increased.

5. The number of groups in the variable and the degrees of freedom within.

6. An HSD determines the difference between two means that is necessary in order for them to be considered significantly different.

Multiple-Choice Questions

1. Using the following mean table and an HSD value = 4, how many groups are significantly different from the first?

Group	1	2	3	4
Mean	4	7	12	13

a. 1

c. 3

b. 2

d. 4

2. Using the following ANOVA source table, compute the HSD at $\alpha = .01$.

	SS	df	MS	F
Between	89	2	44.50	9.77
Within	123	27	4.56	
Total	212	29		

a. 2.62

c. 4.12

b. 2.12

d. 5.08

3. Using the following table of means and an HSD of 3.1, which of the following is true?

Group	1	2	3	4
Mean	8.6	6.8	9.9	11.2

a. The only significant difference is between Groups 2 and 3

b. Group 1 and Group 4 are significantly different

c. Group 2 is significantly different from Group 1 but not from Group 4

d. roup 2 is significantly different from Groups 3 and 4, but not from Group 1

4. A research study is comparing 5 different treatments to a control group. This study has enrolled 120 participants who were randomly assigned to one of the conditions so that there were equal numbers of participants in each group. What is q when computing the HSD for this post hoc test at $\alpha = .05$?

   a. 4.45

   c. 4.87

   b. 4.10

   d. 3.92

5. Using the following table, compute an HSD $\alpha = .05$. (There are equal numbers of participants in each group.)

	SS	df	MS	F
Between	367	4	91.75	5.09
Within	541	30	18.03	
Total	908	34		

   a. 8.10

   c. 6.58

   b. 9

   d. 6.18

6. Using the HSD from the previous question, use the means provided below to determine which of the following statements are true.

Group	1	2	3	4	5
Mean	3.45	9.87	4.6	10.8	11.7

   a. Groups 1 and 3 are significantly different

   b. Group 2 and 4 are not significantly different, but both are significantly different from Group 1

   c. The only significant differences are between Groups 1 and 4 and Groups 1 and 5

   d. None of these groups are significantly different from one another

7. The following are the results of a research trial for a new cholesterol medication. The groups were new medication, placebo group, and control group. If there was a total of 45

participants with equal numbers of participants in each group, then what values would you

use to find q when $\alpha = .05$?

a. 3, 45

c. 3, 42

b. 3, 41

d. 2, 42

8. Using the information from the previous question, using an $MS_{with} = 74.3$, what is the HSD?

a. 6.37

c. 7.66

b. 9.73

d. 8.17

9. Using the following mean table and the information from questions 7 and 8, determine which

groups are significantly different.

Group	1	2	3
Mean	124.5	137.5	140.4

a. Groups 1 and 3 are significantly different

b. Groups 1 and 2 are significantly different

c. Groups 2 and 3 are significantly different

d. A and B

A study was done to determine the influence of domestic violence on depression. Levels of

depression were compared in three groups: physical domestic violence in the home; verbal

domestic violence in the home; and a nonviolence control group. There were 27 total

participants, with equal numbers of participants in each group. Use the following to answer

questions 10 and 11.

	SS	df
Between	748	2
Within	897	24
Total	1645	26

	Physical	Verbal	Control
Mean	14	12	4

253

10. What would the Tukey HSD be for this test at $\alpha = .05$?

    a. 8.05

    b. 7.19

    c. 9.27

    d. 5.95

11. Which groups are significantly different using the HSD test?

    a. Groups 1 and 2 are different from Group 3

    b. Groups 2 and 3 are different from Group 1

    c. Groups 1 and 3 are different from Group 2

    d. All groups are significantly different

Multiple-Choice Answers

1. B

2. A

3. D

4. B

5. C

6. C

7. C

8. C

9. D

10. B

11. A

Module Quiz

1. Conduct an HSD to determine which groups are significantly different at $\alpha = .05$.

	SS	df	MS	F
Between	5,962.67	2	2,981.33	5.9
Within	9,098.29	18	505.46	
Total	15,060.95	20		

	Drug	Control	Placebo
Mean	81.86	42.43	51.57
n	7	7	7

2. Using the tables below, determine which groups are significantly different at α = .05.

	SS	df	MS	F
Between	32	2	16	12.16
Within	75	87	1.32	
Total	107	89		

	1	2	3
Mean	20	34	45
n	30	30	30

3. Using the following ANOVA source table, compute the HSD at α = .05.

	SS	df	MS	F
Between	25	3	8.33	5.21
Within	32	20	1.60	
Total	57	23		

4. A dry cleaner was comparing the effectiveness of three different detergents on removing a stain from a shirt. It found a significant F ratio. The HSD post hoc test revealed an HSD = 1.7. Using the means in the following table, interpret the findings.

	Detergent 1	Detergent 2	Detergent 3
Mean	7.20	9	11.30

5. The following means represent the amounts of hazardous emissions put out by different cars of different car manufactures. If there are equal numbers of cars made by each manufacturer and the HSD = 5.2, which cars are significantly more hazardous than the others in their emissions?

	Manufacturer 1	Manufacturer 2	Manufacturer 3	Manufacturer 4	Manufacturer 5
Mean	12	13	10	8	21

Quiz Answers

1. $3.61\sqrt{\dfrac{505.46}{7}} = 30.67$. The drug group is significantly different from the control group.

2. $3.38\sqrt{\dfrac{1.32}{30}} = 0.71$. All of the groups are significantly different from one another.

3. $3.96\sqrt{\dfrac{1.60}{6}} = 2.04$.

4. Detergent 3 is appears to be the most effective detergent because it is significantly better than Detergent 1.

5. Manufacturer 5 is significantly more hazardous in emissions output any of the other manufacturers. The rest of the manufacturers are not significantly different from one another.

Module 27

Learning Objectives

- Distinguish between the information provided by an F test and a Scheffé test

- Know when it is appropriate to calculate a Scheffé test

- Calculate a Scheffé test

- Report results in APA format

Module Summary

- The Tukey HSD is just one of the many types of post hoc test. A second type of post hoc test is the **Scheffé test**. The Scheffé test creates a series of two-group ANOVAs in which each treatment is compared to all other treatments. The formula for the Scheffé test is as follows:

$$F_{\text{Factor1 vs. Factor2}} = \frac{MS_{\text{Between Factor1 and Factor2}}}{MS_{\text{within}}}$$

- In order to find the numerator for a Scheffé test, you must first recalculate the MS for each pair. This is done by using a slight variation of the original formula for the between-groups SS. Rather than using the SS of all three groups, we use the SS for each individual group for each Scheffé test. Once this new SS has been obtained, it is divided by the df_{bet} from the original ANOVA to obtain the MS for the two groups. Using the original df_{bet} keeps the Type 1 error low, at only a portion of the total error rate.

- The final step in a Scheffé test is determining if there is a significant difference between groups. To do this, this we obtain an F ratio by dividing the MS by the df, but again we use the df for the original F test to maintain a low error level. If the F ratio from the Scheffé test exceeds the F critical value, then the means are significantly different.

257

Computational Exercises

1. Using the following data, determine which groups are significantly different using a Scheffé test at $\alpha = .05$.

Group 1	Group 2	Group 3
50	43	69
55	44	53
58	40	62
57	58	70
45	53	68
45	55	71

	SS	df	MS_{bet}	F
Between	954.33	2	477.17	10.47
Within	683.67	15	45.58	
Total	1638	17		

2. Using the following data, determine which groups are significantly different using a Scheffé test at $\alpha = .05$.

Group 1	Group 2	Group 3
2	14	11
3	2	23
7	6	24
10	0	20
5	0	19
14	9	17
5	10	11

	SS	df	MS_{bet}	F
Between	634.38	2	317.19	12.76
Within	447.43	18	24.86	
Total	1081.81	20		

3. Using the following data, determine which groups are significantly different using a Scheffé test at $\alpha = .05$.

Group 1	Group 2	Group 3
19	19	7
26	26	7
23	28	14
11	26	8
27	24	15
10	15	6
21	26	14
15	21	10
23	28	10
12	16	8

	SS	df	MS_{bet}	F
Between	880.27	2	440.13	17.79
Within	667.90	27	24.74	
Total	1548.17	29		

4. Using the following data, determine which groups are significantly different using a Scheffé test at $\alpha = .05$.

Group 1	Group 2	Group 3	Group 4
13	18	25	26
14	11	35	22
11	10	36	25
10	15	34	28
12	11	36	24
20	20	30	21
16	17	33	22
20	18	28	18
18	14	34	27
19	16	32	16

	SS	df	MS_{bet}	F
Between	1953.02	3	651.01	33.21
Within	823.44	42	19.61	
Total	2776.46	45		

5. Determine which groups are significantly different at $\alpha = .05$ using a Scheffé test.

Group 1	Group 2	Group 3
24	25	25
22	18	19
26	16	15
15	27	22
19	21	19
24	15	20
29	16	20
15	23	20
26	26	16
17	26	19

Computational Answers

1.

	SS	df	MS_{bet}	F
1 & 2	24.08	2	12.04	0.26
1 & 3	574.08	2	287.04	6.30
2 & 3	833.33	2	416.67	9.14

Groups 1 and 2 are significantly different from Group 3.

2.

	SS	df	MS_{bet}	F
1 & 2	1.79	2	0.89	0.04
1 & 3	445.79	2	222.89	8.97
2 & 3	504	2	252	10.14

Group 3 is significantly different from Groups 1 and 2.

3.

	SS	df	MS_{bet}	F
1 & 2	88.20	2	44.10	1.80
1 & 3	387.20	2	193.60	7.91
2 & 3	845	2	422.50	17.27

Groups 1 & 2 are significantly different from Group 3.

4.

	SS	df	MS_{bet}	F
1 & 2	0.45	3	0.15	0.01
1 & 3	1445	3	481.67	24.56
1 & 4	288.80	3	96.27	4.91
2 & 3	1496.45	3	498.82	25.44
2 & 4	312.05	3	104.02	5.30
3 & 4	441.80	3	147.27	7.51

Group 1 and 2 are significantly different from Groups 3 and 4. Also, Groups 3 and 4 are significantly different from each other.

5. This was a trick question, as the F ratio from the original ANOVA is not significant. Therefore, there is no significant difference among the groups.

True/False Questions

1. A Scheffé test determines significant differences by comparing all the possible pairs of means.

2. The MS_{with} for a Scheffé test is obtained from the original ANOVA.

3. The Scheffé test requires a different MS between groups for each comparison.

4. The SS between groups in a Scheffé test is found by dividing up the total SS_{bet} from the original ANOVA.

5. You are less likely to make a Type 1 error with an HSD than with a Scheffé test.

6. A Scheffé test provides one F ratio for determining significant differences.

7. The Scheffé test is more conservative than the HSD test.

8. When calculating the MS_{bet} for a Scheffé test, you should the df_{bet}.

True/False Answers

1. True   2. True   3. True

4. False	6. False	8. True
5. False	7. True	

Short-Answer Questions

1. How do Scheffé and HSD tests differ?

2. Which post hoc test is considered more conservative? Why?

3. Using the formula, how does a Scheffé test maintain the Type 1 error rate?

4. What is the difference between a Scheffé test F test and an ANOVA F test?

5. What are the steps of conducting a Scheffé test?

Answers

1. The Scheffé test compares the amount of variability between two groups (the recalculated SS_{bet}) to the total amount of error variability in the problem. Significance is then determined with an F ratio. An HSD determines the amount by which two means must differ in order to be considered significantly different. This is based on a portion of the error variability and a correction factor (q).

2. A Scheffé test is considered more conservative because it compares the variability between two groups to all of the error variability (as opposed to only the error between the groups). This is also represented in the degrees of freedom that are used.

3. A Scheffé test maintains the Type 1 error rate by using the original df_{with} and the MS_{with} in calculating the F ratio.

4. A Scheffé F test compares just two groups, whereas an ANOVA F test compares all of the groups.

5. (a) Determine the SS_{bet} for the two groups being compared. (b) Determine the MS_{bet} for the two groups by dividing by the original df_{bet}. (c) Divide the MS_{bet} for the two groups by the

MS_{with} to obtain the F ratio. (d) Compare the obtained F ratio to a critical F ratio obtained by using the original df_{bet} and df_{with}.

Multiple-Choice Questions

1. How many groups are compared per F ratio in a Scheffé test?

    a. 2

    b. 3

    c. 4

    d. As many as there in the ANOVA

2. If 5 groups are compared in a single ANOVA, how many F ratios will be computed when doing a Scheffé post hoc test?

    a. 5

    b. 7

    c. 8

    d. 10

3. An ANOVA compared 3 groups and contained a total of 15 participants. When conducting a Scheffé test, how many degrees of freedom should you use when looking up the F critical value?

    a. 2, 14

    b. 2, 12

    c. 3, 15

    d. 3, 12

4. The following table provides the SS of a Scheffé test comparing the amount of completed passes made by 4 football teams. If the MS_{with} for the primary ANOVA was 22,578 and there were 60 total players across the teams, what is the F ratio for the comparison between Teams 2 and 3?

	SS	MS_{bet}
1 & 2	46,728.38	15,576.13
1 & 3	152,960.67	50,986.89
2 & 3	193,788.48	64,596.16

1 & 4	308,493.38	102,831.13
2 & 4	365,461.44	121,820.48
3 & 4	249,573.62	83,191.20

a. 2.86

c. 6.23

b. 3.78

d. 4.53

5. Based on the table in the previous question, which statement is true?

   a. Groups 1 and 2 are significantly different

   b. Group 4 is significantly different from Groups 1 and 2

   c. Group 3 is significantly different from Group 1

   d. Group 4 is significantly different from Groups 3 and 2, but not from Group 1

A study was done to determine the influence of domestic violence on depression. Levels of depression were compared in three groups: physical domestic violence in the home, verbal domestic violence in the home, and a nonviolence control group. There were 27 total participants, with an equal number of participants in each group. Use the following to answer questions 6 and 7.

	SS	df
Between	748	2
Within	897	24
Total	1,645	26

	Physical	Verbal	Control
Mean	14	12	4

6. In doing a Scheffé test, what is the value of MS_{with}?

   a. 37.38

   c. 23.12

   b. 45.78

   d. 6.37

7. Using the following Scheffé table, which groups appear to be significantly different?

	SS	MS_{bet}
1 & 2	222.44	74.15
1 & 3	1,165.21	388.40
2 & 3	20.07	6.69

a. Groups 1 and 2

c. Groups 1 and 3

b. Groups 2 and 3

d. All groups are significantly different.

Multiple-Choice Answers

1. A

5. B

2. D

6. A

3. B

7. C

4. A

Module Quiz

1. Using the raw data below, calculate which groups are significantly different using a Scheffé

   test at $\alpha = .05$.

Drug	Control	Placebo
92	26	15
93	33	53
89	53	55
95	45	22
36	45	84
79	71	42
89	24	90

2. Using the following data, determine which groups are significantly different using a Scheffé

   test at $\alpha = .05$.

Group 1	Group 2	Group 3
67	72	93
50	49	76
68	72	86
26	40	71

74	35	95
67	72	93
50	49	76

	SS	df	MS	F
Between	2,812.93	2	1,406.47	7.94
Within	3,188	18	177.11	
Total	6,000.93	20		

3. For a Scheffé test, how will you know which groups are significantly different? How does this differ from an HSD test?

4. A CEO is comparing the scores on an aptitude test of three different departments in her company. Unfortunately, she will have to fire the department that gets the lowest score at the end of the financial year. If she found a significant F ratio with an $MS_{with} = 623.37$, indicating that one group performed significantly different from another group, which group should she fire? (Use the following table of results. There were 5 people in each department.)

	SS	df	MS_{bet}	F
1 & 2	1960.82	2	980.41	1.57
1 & 3	6181.35	2	3090.68	4.96
2 & 3	9652.02	2	4826.01	7.74

	Department 1	Department 2	Department 3
Mean	89	91	86

5. The following are the results of a Scheffé test that contained an $MS_{with} = 112$, with 3 groups and a total of 60 participants. Which groups appear significantly different with $\alpha = .05$?

	SS	df	MS_{bet}	F
1 & 2	1,297.92	2	648.96	5.79
1 & 3	499.23	2	924.8	8.26
2 & 3	2,170.83	2	1,170.45	10.45

Quiz Answers

1. The only difference is between the drug and control groups. The results are identical to those that would be obtained from the HSD test.

2.

	SS	df	MS_{bet}	F
1 & 2	28.9	2	14.45	0.08
1 & 3	1,849.6	2	924.8	5.22
2 & 3	2,340.9	2	1,170.45	6.61

Groups 2 and 3 are significantly different.

3. For a Scheffé test, you obtain an individual F ratio for each two-group comparison. For an HSD, a single value is obtained, and the difference between all of the means is compared to this value.

4. It appears that the only significant difference is between Departments 2 and 3. The CEO should let Department 3 go, as it scored the lowest and is significantly lower than the top department. Department 1 is not significantly different from Departments 2 and 3.

5. Groups 1 and 3 and Groups 2 and 3 are different from one another.

Module 28

Learning Objectives

- Express a factorial ANOVA design in symbols

- Determine the number of possible main effects in an ANOVA

- Determine the number of possible interaction effects in an ANOVA

- From a graph of cell means, determine whether or not there are probable main effects or interaction effects

- From a table of cell means, determine whether or not there are probable main effects or interaction effects

Module Summary

- ANOVA has two general uses: (1) to examine group differences among three or more groups for a single independent variable and (2) to assess differences among multiple independent variables (regardless of how many groups are associated with each of those IVs). In the second case, the ANOVA is called a *factorial ANOVA* because it has multiple factors (independent variables).

- ANOVAs with two independent variables are referred to as *two-way ANOVAs*. ANOVAs can have many more factors, with the analysis becoming increasingly more complex with the addition of more factors. Two-way ANOVAs are written as the number of groups for the first IV × the number of groups for the second IV. For example, a study with 3 groups for the first IV and 4 groups for the second IV would be expressed as a 3 × 4 ANOVA. Three-way ANOVAs follow the same pattern. For example, if the previous study had two groups for a third IV, it would be expressed as a 3 × 4 × 2 ANOVA.

- Factorial ANOVA enables you to assess numerous hypotheses at once. The first type of hypothesis addresses whether an IV had a significant effect on the DV. These types of hypotheses are referred to as main effects. There is one main effect per independent variable in the study. Main effects state that there is an effect of one IV on the DV, regardless of any other IVs. Assume you are conducting a study on makeup sales and your factors were brand of makeup (3 types) and specific product (3 types). A main effect for brand would indicate that regardless of what type of product, one particular brand is selling more than others. A main effect for product would indicate that regardless of brand, one particular product is selling more than the others. When graphing the means of main effects without a significant interaction, you can expect the lines connecting the means within each factor to be parallel.

- The second type of hypothesis addressed by a factorial ANOVA is referred to as an interaction. Interactions indicate that a score on the DV depends on particular combinations of the first and second IVs. Returning to our makeup example, a significant interaction would indicate that a particular brand/product combination is selling more than any another brand/product combination. When graphing the means of an interaction effect, you can expect the lines created by connecting the means to intersect.

- It should be noted that graphs or tables for determining significant main effects and interactions can be difficult to interpret when the lines are neither perfectly parallel nor clearly crossed. The method for assessing significant differences again uses the F statistics that are calculated for them and will be discussed in the next module.

Computational Exercises

1. Using the following graph, what are the main effects and interactions (if any)?

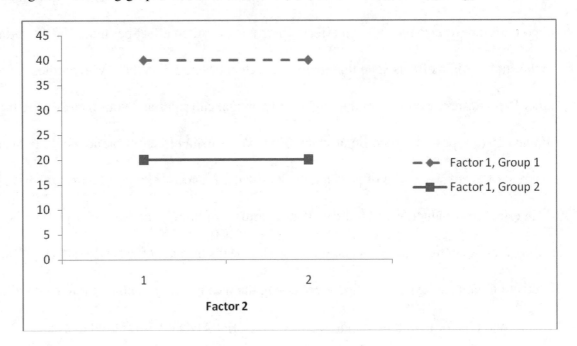

2. Using the following graph, what are the main effects and interactions (if any)?

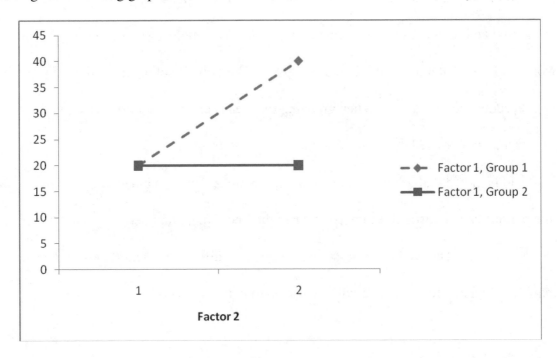

3. Using the following table, what are the main effects and interactions (if any)?

		Factor 2		
		Group 1	Group 2	Group 3
Factor 1	Group 1	10	30	20
	Group 2	10	30	20
	Group 3	10	30	40

4. Using the following table, what are the main effects and interactions (if any)?

		Factor 2	
		Group 1	Group 2
Factor 1	Group 1	45	45
	Group 2	45	45

Computational Answers

1. There is a main effect for Factor 1, but not for Factor 2. There is no interaction.

2. There is a main effect for Factor 1. There is also a main effect for Factor 2. There is also an interaction effect, with those in Factor 1, Group 1, having a mean that is significantly different from those of the other groups.

3. There are two significant main effects and a significant interaction. The third group in Factors 1 and 2 is significantly higher than the rest.

4. There are no significant main effects or interactions in this example.

True/False Questions

1. A study that is examining the cause of hearing damage is investigating the effect of volume of sound and the proximity of the sound to the ear. This would be a factorial ANOVA.

2. An interaction indicates that the effect of one IV on the DV is conditional upon the level of a second IV.

3. A main effect is similar to the results found in a one-way ANOVA.

4. When the row and column means are identical, there is a significant interaction.

5. When graphing the means, overlapping lines indicate a lack of all main effects.

6. When graphing means, separate but parallel lines indicate a main effect for that factor and a lack of an interaction.

7. In real research, you can expect perfectly parallel lines in the absence of an interaction.

8. An ANOVA with three groups for one factor, two groups for the second factor, and four groups for the third factor could be written as a 24-way ANOVA.

9. There are no interactions in the following ANOVA

		Factor 2		
		Group 1	*Group 2*	*Group 3*
	Group 1	10	15	15
Factor 1	Group 2	10	15	15
	Group 3	10	15	15

True/False Answers

1. True

2. True

3. True

4. False

5. False

6. True

7. False

8. False

9. True

Short-Answer Questions

1. When is it appropriate to use a factorial ANOVA?

2. What is a main effect?

3. How does an interaction differ from a main effect?

4. Create a graph for a significant 2×3 interaction effect.

5. Is it possible to create a graph for a three-way interaction? Why or why not?

6. Why is relying solely on a table or graph not the best method for assessing significant differences in a factorial ANOVA?

7. How many effects are possible in a 4-way ANOVA?

8. Where should you inspect on a table of means for a significant interaction effect and a significant main effect?

9. Develop a study for a 3 × 3 ANOVA.

Answers

1. Factorial ANOVAs are used when there are multiple independent variables.

2. A main effect means that there is a significant difference among the group means for a particular factor.

3. An interaction differs from a main effect in that an interaction looks for significant differences using both of the factors. Specifically, it means that the influence of one IV on the DV is dependent upon the influence of a second IV.

4. Answers will vary; however, there should be an indication of non-parallel lines.

5. It is not possible to graph such an interaction in two dimensions. This is because you have no way of showing the effect of the third IV. Alternative methods including creating a 3D graph or creating multiple graphs. The second method entails creating separate graphs (showing the groups of the third IV) and then, within each graph, showing the interactions and main effects of the first two IVs.

6. It is expected that there will be some variation among the group means, even though this difference is not significant. As a result, this difference may lead you to falsely conclude that there is a significant main effect or interaction.

7. There are possibly 4 separate main effects (1 for each IV), with 11 possible interaction terms.

8. A significant interaction will be located within the cell means, and a significant main effect will be located in the row and column means.

9. Answers will vary.

Multiple-Choice Questions

1. In the following table, how many effects are there?

		Factor 2	
		Group 1	Group 2
Factor 1	Group 1	10	15
	Group 2	10	15

    a. Two main effects

    b. A main effect for Factor 1 but not

        Factor 2

    c. Two main effects and an

        interaction

    d. An interaction

2. What would X need to be for there to be a main effect for Factor 2 and no interaction?

		Factor 2	
		Group 1	Group 2
Factor 1	Group 1	10	45
	Group 2	25	X

    a. 45

    b. 20

    c. 30

    d. 55

3. What would X need to be in order for there to be no main effects?

		Factor 2	
		Group 1	Group 2
Factor 1	Group 1	5	30
	Group 2	30	X

    a. 30

    b. 5

    c. 10

    d. 25

4. If all the row and column means of a mean table are the same, then what is significant?

    a. The main effects, but not the interaction

    b. The main effects and the interaction

    c. Possibly the interaction, but not the main effects

    d. At least one main effect

5. In the following graph, how many interactions and main effects are there?

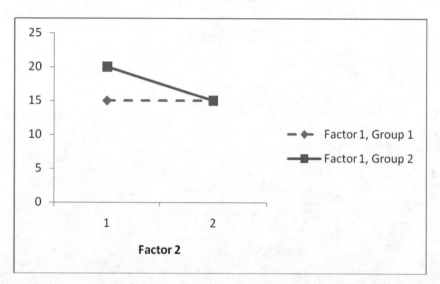

a. Both main effects, but not the interaction

b. Both main effects and the interaction

c. The interaction but neither main effect

d. One main effect and the interaction

Multiple-Choice Answers

1. B
2. D
3. B
4. C
5. D

Module Quiz

1. How many main effects and interactions are present in the following table?

		Factor 2	
		Group 1	Group 2
Factor 1	Group 1	50	40
	Group 2	90	80

2. How many main effects and interactions are present in the following graph?

275

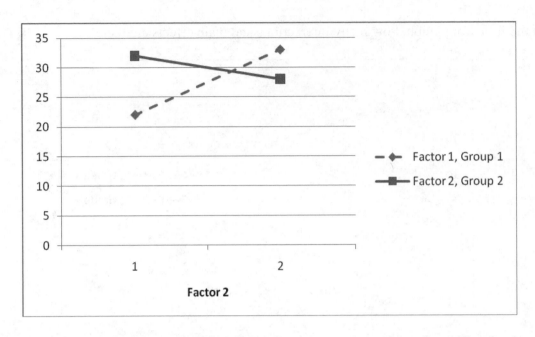

3. A study was conducted on the influence of supervision on improving the skills of a therapist. The study was interested in determining how the theoretical orientation of the supervisor and the length of time of the supervision (1 week, 2 weeks, 3 weeks, 4 weeks) influenced therapist improvement. Here is a table of the results.

		Time in Weeks			
		1	*2*	*3*	*4*
Orientation	Cognitive Behavioral Therapy	20	34	47	45
	Interpersonal	21	29	41	44

Based on the previous table, which effects appear to be significant?

4. When graphing means, what do two parallel (or close to parallel) lines indicate?

5. In a study of career enjoyment, a researcher is interested in determining how people feel about their different careers. They hypothesize that the length of time in which people are involved with a certain career influences their satisfaction. Due to the gender gap in salary and promotion, the researcher also wants to incorporate gender into the analysis. How many IVs are in this study?

Quiz Answers

1. There are two main effects. The main effect for Factor 1 is small, however, so it might or might not be statistically significant.

2. There are two main effects and an interaction. Both main effects are small, so they might or might not be statistically significant.

3. There is one main effect in the table, for time.

4. This indicates that there is no interaction effect among the factors.

5. There are three IVs: Career type, Time (length of career), and Gender

Module 29

Learning Objectives

- Understand the similar logic underlying various test statistics

- Determine the degrees of freedom

- Calculate a factorial ANOVA

- Use a table to interpret F

- Present results in an ANOVA table

- Report results in APA format

Module Summary

- When calculating a factorial ANOVA from raw data, it is important to be organized because of the large number of calculations.

- The first step is to calculate the cell means. A cell mean refers to the average score of those in a particular group combination (such as Group 1 of Factor 1 and Group 1 of Factor 2). Next, the row and column means should be calculated. These are the means for the overall groups in a factor. This indicates that the row mean for Factor 1 would be the mean of all the scores in Factor 1, regardless of the Group on factor 2. The cell means are related to *interaction effects*, whereas the row and column means are related to the main effects.

- The SSs for a factorial ANOVA are calculated in a manner very similar to that of the one-way ANOVA. The purpose of finding the SS is to partition it into two different sections: variability related to differences between the groups and variability related to random error (within). The variability between the groups can be further broken down into variability related to differences on Factor 1, variability related to differences on Factor 2, and variability related to the interaction. The formulas for the total SS are

$$\text{Deviation Method: } SS_{\text{tot}} = \sum_{1}^{N}(X - M_{\text{tot}})^2$$

$$\text{Raw Score Method: } SS_{\text{tot}} = \sum_{1}^{N}X^2 - \frac{(\sum[X_{\text{tot}}])^2}{N}$$

The formulas for the within SS are

$$\text{Deviation Method: } SS_{\text{with}} = \sum_{1}^{k}\sum_{1}^{n}(X - M_{\text{cell}})^2$$

$$\text{Raw Score Method: } SS_{\text{with}} = \sum_{1}^{N}X^2 - \sum_{1}^{k}\frac{(\sum[X_{\text{cell}}])^2}{n_{\text{cell}}}$$

The formulas for the between SS are as follows. Note that formulas for additional factors are the same.

$$\text{Factor 1, Deviation Method: } SS_{\text{Factor1}} = \sum_{1}^{k}(M_{\text{Factor1}} - M_{\text{tot}})^2$$

$$\text{Factor 1, Deviation Method: } SS_{\text{Factpr1}} = \sum_{1}^{N}\frac{(\sum[X_{\text{cell}}])^2}{n_{\text{Factor1}}} - \frac{(\sum[X_{\text{tot}}])^2}{N}$$

The formulas for the interaction term are as follows:

$$SS_{\text{interaction}} = SS_{\text{tot}} - (SS_{\text{with}} + SS_{\text{Factor 1}} + SS_{\text{Factor 2}})$$

- After the SS has been partitioned, the next step is to calculate the appropriate degrees of freedom. Remember that the degrees of freedom are used in calculating the mean squares.

$$df_{\text{tot}} = N - 1$$

$$df_{\text{with}} = (\text{\# groups in Factor 1})(\text{\# groups in Factor 2})(n_{\text{cell}} - 1)$$

$$df_{\text{Factor 1}} = (\text{\# groups in Factor 1}) - 1$$

$$df_{\text{Factor 2}} = (\text{\# groups in Factor 2}) - 1$$

$$df_{\text{interaction}} = [(\text{\# groups in Factor 1}) - 1] \, [(\text{\# groups in Factor 2}) - 1]$$

- The next step is to calculate the *Mean Square*. The formulas used are similar to that of the one-way ANOVA.

$$MS_{with} = \frac{SS_{with}}{df_{with}} \; ; MS_{Factor1} = \frac{SS_{Factor1}}{df_{Factor1}} \; ; MS_{Factor2} = \frac{SS_{Factor2}}{df_{Factor2}}$$

$$MS_{Interaction} = \frac{SS_{Interaction}}{df_{Interaction}}$$

- The final step is to calculate the *F* ratios. In a factorial ANOVA, there are multiple *F* ratios, one for each main effect and one for the interaction term.

$$F_{Factor1} = \frac{MS_{Factor1}}{.MS_{within}} \; ; F_{Factor2} = \frac{MS_{Factor2}}{MS_{within}} \; ; F_{Interaction} = \frac{MS_{Interaction}}{MS_{Within}}$$

- After obtaining the *F* ratios, the next step is to determine if they are significant (that is, are the group differences a lot or a little?). This is done in a similar manner to that of a one-way ANOVA by using the table found in Appendix D. Because you have obtained three *F* ratios, you will need to find three different critical values.

- Similar to a one-way ANOVA, the results can be displayed easily in a source table. Notice that in a factorial ANOVA, there are additional rows for the interaction effect(s).

Source	SS	df	MS	F
Factor 1	$SS_{Factor\ 1}$	$df_{Factor\ 1}$	$SS_{Factor\ 1}/df_{Factor\ 1}$	$MS_{Factor\ 1}/MS_{with}$
Factor 2	$SS_{Factor\ 2}$	$df_{Factor\ 2}$	$SS_{Factor\ 2}/df_{Factor\ 2}$	$MS_{Factor\ 2}/MS_{with}$
Interaction	$SS_{nteraction}$	$df_{interaction}$	$SS_{interaction}/df_{Interaction}$	$MS_{interaction}/MS_{interaction}$
Within	SS_{with}	df_{with}	SS_{with}/df_{with}	
Total	SS_{tot}	df_{tot}		

Computational Exercises

1. Partition the following the *SS* of the following groups.

	Factor 1	
Factor 2 Group	Group 1	Group 2
1	61	62
1	15	49
1	72	12
1	53	9
2	49	31
2	81	85
2	1	73
2	20	74

2. Partition the *SS* of the following groups.

	Factor 1		
Factor 2 Group	Group 1	Group 2	Group 3
1	76	46	69
1	64	79	45
1	44	42	77
1	75	42	45
2	68	43	56
2	60	69	53
2	50	60	48
2	40	44	72
3	66	56	73
3	66	74	61
3	55	63	40
3	69	76	66

3. A study examined a new treatment for social anxiety by comparing the new treatment to an established treatment and a separate control group (three groups in total). After the study has been conducted, the therapists anecdotally report that they noticed that females seemed to be more responsive to the treatment than males. Here are the data from the study's primary outcome measure, which is the treatment. Now add the gender variable and calculate the factorial ANOVA.

	Treatment		
Gender	New Treatment	Established Treatment	Control
Male	2	9	14
Male	6	8	19
Male	3	5	16
Male	8	3	22
Male	7	3	24
Female	5	9	19
Female	7	7	25
Female	5	5	10
Female	5	12	15
Female	9	7	17

a. What are the null and alternative hypotheses?

b. What are the main effects and interactions (if any) at the .05 error level?

c. Present your findings in APA format.

4. Within the diagnosis of schizophrenia, there are multiple subtypes. A new research study seeks to compare the effectiveness of four separate treatments for schizophrenia (SSRI, antipsychotic, therapy, and control) across two of the subtypes (primarily showing positive symptoms and primarily showing negative symptoms). Using the data obtained below that consist of an observer's judgment of change in the client, calculate the factorial ANOVA.

	Treatment			
Subtype	SSRI	Antipsychotic	Therapy	Control
Positive	1	5	13	13
Positive	10	7	12	19
Positive	1	5	10	19
Positive	1	6	18	17
Positive	10	10	20	9
Positive	6	6	13	11
Negative	5	8	19	20
Negative	10	6	7	15
Negative	4	8	20	7
Negative	6	4	12	7
Negative	9	2	9	10
Negative	8	5	15	11

a. What are the null and alternative hypotheses?

b. What are the main effects and interactions (if any) at the .05 error level?

c. Present your findings in APA format.

5. You are hired by an ice cream maker to develop a new ice cream flavor. Ice cream consists of a base flavor (the actual ice cream) and candy that is mixed in. You decide to assess chocolate and vanilla as your base flavors. You include chocolate chips and peanut butter cups as your mix-ins. You give a group of $N = 108$ people one of each of the 4 combinations and rank their opinion of the flavor they tasted on a scale of 1-10. The company wants to know what combinations you could recommend based on your research, with an error level of .05.

	Vanilla	Chocolate	Flavor Mean
Choc Chips	5.1	6.2	5.65
PB Cups	9.76	6.21	7.99
Mix-in Mean	7.43	6.21	6.82

Source	SS	df	MS	F
	450			
	125			
	80			
	785			

a. Complete the above source table.

b. What are the main effects and interactions (if any) at the .01 error level?

c. Present your findings in APA format.

6. A jeweler is curious as to what women prefer with regard to their engagement rings. She decides to obtain ratings (1-10) from 32 different women (an equal number in each group) on their opinion of wedding rings in terms of the shape of the stone and the ring setting. She

finds that the most popular shapes are princess and pear, and the most popular settings are gold and platinum. Here are the raw scores of satisfaction she obtained from the women.

	Setting	
Shape	Gold Setting	Platinum Setting
Princess	6	8
Princess	10	3
Princess	10	9
Princess	10	5
Princess	8	4
Princess	10	7
Princess	7	7
Princess	9	3
Pear	10	5
Pear	3	6
Pear	9	7
Pear	9	6
Pear	10	5
Pear	9	10
Pear	6	5
Pear	7	3

a. What are the null and alternative hypotheses?

b. What are the main effects and interactions (if any) at the .05 error level?

c. Present your findings in APA format.

Computational Answers

1. $SS_{tot} = 1187.44$; $SS_{Factor 1} = 115.56$; $SS_{Factor 2} = 410.06$; $SS_{interaction} = 2047.56$; $SS_{with} = 9314.25$.

2. $SS_{tot} = 5560.22$; $SS_{Factor 1} = 67.39$; $SS_{Factor 2} = 439.06$; $SS_{interaction} = 375.78$; $SS_{with} = 4678$.

3. a. Null: Treatment and gender are not related to changes anxiety scores.

   Alternative: Treatment and gender are related to changes in anxiety scores.

   b.

	SS	df	MS	F	Sig.
Treatment	942.20	2	471.10	39.31	< .05
Gender	2.13	1	2.13	0.18	> .05
Interaction	22.87	2	11.44	0.95	> .05
Within	287.60	24	11.98		
Total	1254.80	29			

There is a significant main effect for treatment; however, there is no main effect for gender and there is no significant interaction.

c. There is a significant main effect for treatment, $F(2, 24) = 39.31$, $p < .05$. There is no significant main effect for gender, $F(1, 24) = 0.18$, $p > .05$. There is no significant interaction, $F(2, 24) = 0.95$, $p > .05$.

4. a. Null: Treatment and subtype do not affect change in schizophrenia.

Alternative: Treatment and subtype do affect change in schizophrenia.

b.

	SS	df	MS	F	Sig
Treatment	701.87	3	233.96	15.52	< .05
Subtype	4.69	1	4.69	0.31	> .05
Interaction	40.73	3	13.58	0.90	> .05
Within	603.17	40	15.08		
Total	1350.48	47			

There is a significant main effect for treatment, but not for subtype. There was no significant interaction.

c. There is a significant main effect for treatment modality, $F(3, 40) = 15.52$, $p < .05$; however, there was no main effect for subtype, $F(1, 40) = 0.31$, $p > .05$. There was also no significant interaction, $F(3, 40) = 0.90$, $p > .05$.

5. a.

	SS	df	MS	F
Between treatments	450	3		
Flavor	125	1	125	38.81
Mix-in	80	1	80	24.84
Flavor × Mix-in	245	1	245	76.06
Within treatments	335	104	3.2	
Total	785	107		

b. There are two significant main effects and a significant interaction.

c. There was a significant main effect for base flavor, $F(1, 104) = 38.81$, $p < .05$. There was also a significant main effect for mix-in, $F(1, 104) = 24.84$, $p < 05$. These main effects were qualified by a significant interaction, $F(1, 104) = 76.06$, $p < .05$.

The interaction suggested that the most preferred combination was vanilla ice cream with peanut butter cups mixed in.

6.   a. Null: Preference does not differ based setting or shape.

      Alternative: Preference does differ based setting or shape.

   b.

	SS	df	MS	F	Sig
Setting	50	1	50	11.22	< .05
Shape	1.13	1	1.13	0.25	> .05
Interaction	2	1	2	0.45	> .05
Within	124.75	28	4.46		
Total	177.88	31			

c. There was a significant main effect for setting, $F(1, 28) = 11.22$, $p < .05$. There was not a significant main effect for shape, $F(1, 28) = 0.25$, $p > .05$. Finally, there was no significant interaction, $F(1, 28) = 0.45$, $p > .05$.

True/False Questions

1. In a factorial ANOVA, all of the between-treatment variabilities are compared to the same random error component.

2. The degrees of freedom used to determine the critical value can change depending on which main effect and interaction you are assessing.

3. A 2×2 factorial ANOVA has 32 participants in total (with an equal amount of participants in each group). The correct degrees of freedom for the interaction are 1 and 29.

286

4. A significant interaction may indicate that the highest group mean is different from that suggested merely by the main effects.

5. In a 2 × 2 factorial ANOVA with no significant interaction but two significant main effects, you would need to do two post hoc tests to determine the significant differences among the main effects.

6. In a 3 × 3 × 2 ANOVA, the within subject variability is divided into four sections.

True/False Answers

1. True	3. False	5. False
2. True	4. True	6. False

Short-Answer Questions

1. In a 3 way ANOVA, how many F ratios should you expected to calculate? What are they?

2. In a 3 × 6 ANOVA, how many different groups can you expect to compare in the main effects and the interaction?

3. The formula for an interaction SS can be thought of as $SS_{\text{tot between variability}} - SS_{\text{Factor 1}} - SS_{\text{Factor 2}}$ = $SS_{\text{interaction}}$. How does this formula conceptualize the variability associated with an interaction?

4. An ANOVA consists of three F tests, each with its own critical value. Does this mean that an ANOVA is increasing the chance of committing a Type 1 error?

5. Develop a study for a 3 × 3 × 2 ANOVA.

Answers

1. There will be 7 F ratios: 3 main effects and 4 interactions.

2. The first main effect would compare 3 groups. The second main effect would compare 6 groups. The interaction would compare 18 groups.

3. This formula suggests that the variability (*SS*) for an interaction is attributed to differences among the groups that is not better accounted for by the factors. In other words, it is between-treatment variability that is left over after the variability for the factors has been determined.

4. No, this is not the case. This is because of the shape of the *F* distribution and the number of degrees of freedom that are used in determining the critical value.

5. Answers will vary.

Multiple-Choice Questions

1. A 2×2 study contains $N = 40$ people evenly distributed across the groups. What are the degrees of freedom for this study?

   a. $df_{tot} = 39$; $df_{Factor\ 1} = 2$; $df_{Factor\ 2} = 2$; $df_{interaction} = 4$; $df_{with} = 31$

   b. $df_{tot} = 39$; $df_{Factor\ 1} = 1$; $df_{Factor\ 2} = 1$; $df_{interaction} = 2$; $df_{with} = 35$

   c. $df_{tot} = 39$; $df_{Factor\ 1} = 1$; $df_{Factor\ 2} = 1$; $df_{interaction} = 1$; $df_{with} = 36$

   d. $df_{tot} = 40$; $df_{Factor\ 1} = 1$; $df_{Factor\ 2} = 1$; $df_{interaction} = 1$; $df_{with} = 36$

2. In the following table, which effects are significant at an error level of .05?

	SS	*df*	*MS*	*F*
Factor 1	25	1	25	1.76
Factor 2	12	1	12	0.85
Interaction	14	1	14	0.99
Within	213	15	14.20	
Total	264	18		

   a. The main effects, but not the interaction

   b. The main effects and the interaction

   c. The interaction but not the main effects

   d. None are significant

3. For the main effect for Factor 1 in the previous table, which of the following is the correct presentation of the result in APA format?

   a. $F(1, 15) = 1.76, p < .05$ c. $F(1, 15) = 1.76, p > .05$

   b. $F(1, 18) = 1.76, p < .05$ d. $F(1, 18) = 1.76, p > .05$

4. Using the following table, which F ratio(s) are significant at an error level of .05?

	SS	df	MS
Factor 1	32	1	32
Factor 2	123	4	30.75
Interaction	245	4	61.25
Within	310	35	8.86
Total	710	44	

   a. Factor 1   c. The interaction

   b. Factor 2   d. All of the F ratios are significant

5. Using the following table, what are the *Mean Squares*?

	SS	df
Factor 1	450	2
Factor 2	345	3
Interaction	987	2
Within	1247	18
Total	3029	25

   a. $MS_{Factor\ 1} = 225$; $MS_{Factor\ 2} = 115$; $MS_{interaction} = 493.5$; $MS_{with} = 69.28$

   b. $MS_{Factor\ 1} = 50$; $MS_{Factor\ 2} = 65$; $MS_{interaction} = 214.5$; $MS_{with} = 632$

   c. $MS_{Factor\ 1} = 127$; $MS_{Factor\ 2} = 3214$; $MS_{interaction} = 1246$; $MS_{with} = 5434$

   d. $MS_{Factor\ 1} = 324$; $MS_{Factor\ 2} = 654$; $MS_{interaction} = 324$; $MS_{with} = 987$

6. Using the table in question 5, which F ratio(s) are significant?

   a. Factor 1   c. The interaction

   b. Factor 2   d. All of the F ratios are significant

7. Without altering the *SS* or the Type 1 error level, how could a researcher increase the chances that he or she would obtain a significant result?

   a. Include additional IVs

   b. Increase the sample size

   c. Select another sample

   d. There is nothing the researcher could do

8. How many pieces will the between subject variability be divided into in a 3 × 4 ANOVA? In other words, how many "between" rows will there be in the ANOVA summary (or source) table?

   a. 3

   b. 4

   c. 12

   d. 7

Multiple-Choice Answers

1. C

2. D

3. C

4. C

5. A

6. C

7. B

8. A

Module Quiz

1. Two music artists have recently left their respective groups to pursue solo careers. Their record label is concerned about how their fans will perceive this switch and ask 4 groups of 10 people each their opinion on either one of the groups or the prospect of seeing either artist in a solo career (no one is asked more than one question for fear that rumors might spread). Preference is rated on a 1-20 scale. Here are the data they obtain.

	Solo	Group	Means
Artist A	10.8	6.3	8.55
Artist B	5.4	6.1	5.80
Means	8.1	6.2	

Source	SS	df	MS	F
Between treatments	327			
Solo or Group	89			
Artist	122			
Solo × Artist				
Within treatments				
Total	789			

a. Complete the above source table.

b. What are the main effects and interactions (if any) at the .01 error level?

c. Present your findings in APA format.

2. The jeweler from Computational Exercise 6 has just received a large shipment of emeralds, rubies, and sapphires. She has the opportunity to turn any of these stones into earrings, rings, or bracelets. She wants to ensure that she sells the highest possible number of the pieces she creates. As a result, she reviews the number of sales of each combination of stone and piece from $N = 45$ stores (an equal number of stores per group) by different nearby jewelers. Here are the data:

		Stone	
Shape	Ruby	Emerald	Sapphire
Earring	11	10	6
Earring	9	21	21
Earring	23	10	15
Earring	18	14	21
Earring	20	7	10
Ring	18	19	21
Ring	5	14	24
Ring	6	10	7
Ring	25	15	24
Ring	21	3	14
Bracelet	22	13	15
Bracelet	13	22	7
Bracelet	24	21	15
Bracelet	11	11	8
Bracelet	8	22	23

a. What are the null and alternative hypotheses?

b. What are the main effects and interactions (if any) at the .05 error level?

c. Present your findings in APA format.

3. You are interested in assessing how different environments affect college drinking. You also hypothesize that having an alcoholic parent will interact with the environment to impact the number of drinks consumed. You conduct a study observing the number of drinks consumed in three different college settings: a dorm room, a house party, and a bar. You are able to successfully observe $N = 30$ college students; $n = 10$ different students are observed in each setting. Half of the sample in each setting sample had an alcoholic parent. You obtain the following descriptive statistics.

	Dorm	House Party	Bar	
Alcoholic Parent	2.4	10.2	12.8	8.47
No Alcoholic Parent	2.1	12.3	15.4	9.93
	2.25	11.25	14.1	

Source	SS	df	MS	F
	587	2		
	70			
	487			
Total	1257			

a. Complete the above source table.

b. What are the main effects and interactions (if any) at the .05 error level?

c. Present your findings in APA format.

4. Three new teachers are interested in determining how well students respond to their different teaching techniques: rewarding, punitive, and experiential. They decide to compare the final grades of the $n = 6$ students in each of their classes to determine which teacher has the most effective technique. One of the teachers suggests that the gender of the student may have an impact on response to the technique. There is an equal number of each sex in each class.

The teachers agree to incorporate gender into their study as well. Here are the final grades that they obtained. Is one teaching technique more effective, and does gender play a role? Use a .05 error level and present your findings in APA format.

Gender	Technique		
	Rewarding	Punitive	Experiential
Male	90	86	87
Male	89	89	87
Male	81	84	92
Male	81	90	92
Male	79	75	94
Male	84	90	92
Female	77	78	90
Female	84	87	92
Female	84	85	90
Female	82	88	95
Female	75	79	98
Female	90	86	98

5. In conducting research on treatment, it is common to use multiple therapists; however, this could be a potential problem, as different therapists may have varying levels of effectiveness. A research study on depression is comparing interpersonal therapy and cognitive behavioral therapy. It is using two therapists, who want to ensure that participants are responding to both therapists in a similar manner. Using the following data, determine if there is any difference between the therapists and how (if at all) it affects the type of therapy used to treat depression. The scores represent a self-report measure of depression.

Therapist	Treatment	
	Interpersonal	Cognitive-Behavioral
Therapist A	17	8
Therapist A	6	5
Therapist A	10	20
Therapist A	23	12
Therapist A	17	12
Therapist A	6	13
Therapist B	20	23
Therapist B	17	17
Therapist B	25	11
Therapist B	16	10
Therapist B	5	13
Therapist B	17	15

Quiz Answers

1. a.

Source	SS	df	MS	F
Between treatments	327	3		
Solo or Group	89	1	89	6.94
Artist	122	1	122	9.51
Solo × Artist	116	1	116	9.04
Within treatments	462	36	12.83	
Total	789	39		

b. There are two main effects and one significant interaction.

c. There was a significant main effect for individual solo or group, $F(1, 36) = 6.94$, $p < .01$. There was also a significant main effect for artist, $F(1, 36) = 9.51, p < .01$.

These main effects were qualified by a significant interaction, $F(1, 36) = 9.04, p < .01$.

2. a. Null: There is no difference in preference, regardless of piece or stone.

Alternative: There is a difference in preference depending on piece or stone.

b.

	SS	df	MS	F	Sig
.Stone	18.98	2	9.49	0.21	> .05
Piece	12.04	2	6.02	0.13	> .05
Interaction	145.69	4	36.42	0.82	> .05
Within	1607.20	36	44.64		
Total	1783.91	44			

c. There was no significant main effect for stone, $F(2, 36) = 0.21, p > .05$. There was no significant main effect for piece, $F(2, 36) = 0.13, p > .05$. There was no significant interaction, $F(4, 36) = 0.82, p > .05$.

3. a.

Source	SS	df	MS	F
Environment	587	2	293.5	14.46
Alcoholic Parent	70	1	70	3.45
Environment × AP	113	2	56.5	2.78
Within Treatments	487	24	20.29	
Total	1257	29		

b. There is one significant main effect, for environment.

c. There was a significant main effect for environment, $F(2, 24) = 14.46, p < .05$. There was not a significant main effect for having an alcoholic parent, $F(1, 24) = 3.45, p > .05$. There was not a significant interaction, $F(2, 24) = 2.78, p > .05$.

4.

	SS	df	MS	F	Sig.
Type	579.50	2	289.75	14.06	< .05
Gender	0.44	1	0.44	0.02	> .05
Interaction	51.72	2	25.86	1.26	> .05
Within	618.33	30	20.61		
Total	1250	35			

There is a significant main effect for teaching subtype, $F(2, 30) = 14.06, p < .05$. There was no significant main effect for gender, $F(1, 30) = .02, p > .05$. There was also no significant interaction, $F(2, 30) = 1.26, p > .05$. It appears that experiential teaching is the most effective method, regardless of gender.

5.

	SS	df	MS	F	Sig.
Therapy	16.67	1	16.67	0.48	> .05
Therapist	66.67	1	66.67	1.91	> .05
Interaction	0.17	1	0.17	0	< .05
Within	698.33	20	34.92		
Total	781.83	23			

There was no significant main effect for therapy type, $F(1, 20) = 0.48, p > .05$. There was also no significant main effect for therapist, $F(1, 20) = 1.91, p > .05$. There was a significant interaction, $F(1, 20) < 0.01, p < .05$. It appears that both therapies were equally effective and that both therapists performed in a similar manner.

Module 30

Learning Objectives

- Understand the similar logic underlying various test statistics

- Determine the degrees of freedom

- Calculate a one-variable chi-square

- Use a table to interpret χ^2

- Report results in APA format

Module Summary

- Tests that are grounded in the probabilities for populations are referred to as *parametric tests*.
  In certain cases such as using small samples, dealing with highly skewed data, and dealing
  with data that are not based on an interval scale, it may be more useful to use a
  nonparametric test. *Nonparametric tests* are not based on population parameters (such as
  population means or standard deviations) and are thus able to navigate the previously
  mentioned situations.

- A *chi-square* test assesses the statistical significance of frequencies in different nominal
  categories. In other words, a chi-square test assesses if one group is similar to another group.
  As a result, a chi-square test is also referred to as a *goodness-of-fit test*, because it tests
  whether the frequencies in one group "fit" the expected frequencies.

- The logic of the chi-square tests follows the logic of all the previous hypothesis tests. It
  compares the differences between the obtained group frequencies and the expected
  frequencies. Because there is no population standard deviation, the denominator is an
  expected frequency rather than a standardized error term. The formula for a chi-square test is
  as follows:

$$\chi^2 = \sum \left[\frac{(f_0 - f_E)^2}{f_E} \right]$$

- The difficult part of a chi-square test is obtaining the correct expected frequencies. For this section, the expected frequencies will be provided. However, in a real-world situation, it can be difficult to ground your expectations of frequencies in empiricism.

- After obtaining your chi-square value, use the table found in Appendix F to determine if the difference between the expected and obtained frequencies is significant. The chi-square test can only be one-tailed, as is reflected by the table. This is because squaring ensures that a chi-square statistic is always a positive value. The df for a chi-square test is equal to the number of categories in the IV – 1.

- Chi-squares are expressed in APA format as the following: $\chi^2(df, n)$ = chi-square value, p value; for example, $\chi^2(3, n = 10) = 13.21$, $p < .05$.

Computational Exercises

(NOTE: You may round percentages and calculated values to whole numbers in these exercises. If you do not round, your answers will vary slightly from the given answers, but not by much.)

1. A recent nationwide poll suggested that 28% of the nation supports a specific politician. Your local town appears to have a different political ideology from that of the nation, and you want to test this hypothesis. You are able to poll 400 members of your town and find that 132 support the politician.

    a. What are the hypotheses?

    b. Create a table that shows your comparison.

    c. Calculate χ^2.

    d. Interpret your results at a .05 error level.

2. A college has a 7-to-9 male-to-female ratio. The incoming class has 432 males and 520 females. Does this incoming class differ from the college overall?

   a. What are the hypotheses?

   b. Create a table that shows your comparison.

   c. Calculate χ^2.

   d. Interpret your results at a .05 error level.

3. A department store divides its sales year into four seasons: spring, summer, fall, and winter. It usually sells 300 items in the spring and again in the summer, 400 in the fall, and 550 in the winter. This year, the store has a surprisingly good year and sells 310 in the spring, 330 in the summer, 453 in the fall, and 574 in the winter. Would you classify this as a different seasonal sales pattern for the store?

   a. What are the hypotheses?

   b. Create a table that shows your comparison.

   c. Calculate χ^2.

   d. Interpret your results at a .05 error level.

4. Data on income across the nation was recently released. 20% earn less $20,000 per year, 13% earn between $20,000 and 40,000 32% earn between $40,000 and $60,000, 15% earn between $60,000 and $80,000, 10% earn between $80,000 and $100,000, 5% earn between $100,000 and $200,000, and 5% earn more than $200,000. A major city claims that it is richer than average, having 100,000 residents that fall in different income brackets as follows: 13% earn less than $20,000 per year, 15% earn between $20,000 and $40,000, 16% earn between $40,000 and $60,000, 20% earn between $60,000 and $80,000, 20% earn

between $80,000 and $100,000, 10% earn between $100,000 and $200,000, and finally, 6% earn more than $200,000. Is the city correct in its claim?

    a. What are the hypotheses?

    b. Create a table that shows your comparison.

    c. Calculate χ^2.

    d. Interpret your results at a .05 error level.

5. A songwriter receives a complaint from a guitarist that he has written a song that is too complex to play. The guitarist states that he can normally play a song 85% correctly on his first attempt. With the new song, he could correctly play only 329 of the 813 notes in the song. Is the guitarist's complaint warranted?

    a. What are the hypotheses?

    b. Create a table that shows your comparison.

    c. Calculate χ^2.

    d. Interpret your results at a .01 error level.

Computational Answers

1a. Null: Your town does not hold a different opinion from that of the nation.

    Alternative: Your town holds a different opinion from that of the nation.

1b.

Support		Does not Support	
Observed	Expected	Observed	Expected
132	112	268	288

1c. $\chi^2 = \dfrac{(132-112)^2}{112} + \dfrac{(268-288)^2}{288} = 4.96.$

1d. Reject the null. Your town has a higher level of support for the politician than expected.

2a. Null: The incoming class has a similar gender ratio to that of the college.

Alternative: The incoming class has a different gender ratio from that of the college.

2b.

	Male		Female	
	Observed	Expected	Observed	Expected
	432	416	520	536

2c. $\chi^2 = \dfrac{(432-416)^2}{416} + \dfrac{(520-536)^2}{536} = 1.09.$

2d. Critical $\chi^2 = 3.84$. Retain the null. The incoming male-to-female ratio is not different from the ratio at the college.

3a. Null: The sales this year were not different from sales of previous years.

Alternative: The sales this year were different from sales of previous years.

3b.

Spring		Summer		Fall		Summer	
Observed	Expected	Observed	Expected	Observed	Expected	Observed	Expected
310	323	330	323	453	430	574	591

3c. $\chi^2 = 2.39$.

3d. Critical $\chi^2 = 7.81$. Retain the null. This year, the store sold its items at a rate across seasons that was similar to the pattern in previous years.

4a. Null: The income pattern in this town is not significantly different from that of the nation.

Alternative: The income pattern in this town is significantly different from that of the nation.

4b.

Less than 20		20-40		40-60		60-80	
Observed	Expected	Observed	Expected	Observed	Expected	Observed	Expected
13	20	15	13	16	32	20	15

80-100		100-200		Greater than 200	
Observed	Expected	Observed	Expected	Observed	Expected
20	10	10	5	6	5

4c. $\chi^2 = 27.62$.

4d. Critical $\chi^2 = 12.59$. Reject the null. The income of the town is significantly different from that of the nation. Based on the directions of the differences between the observed and expected values, the city does appear to be richer than what we would we expect given the national percentages.

5a. Null: The song is no more or less complex than prior songs.

Alternative: The song is more or less complex than prior songs. .

5b.

Notes Played Correctly		Notes not Played Correctly	
Observed	Expected	Observed	Expected
329	691	484	122

5c. $\chi^2 = 1,263.77$.

5d. Critical $\chi^2 = 3.84$. Reject the null. The song is significantly different in difficulty from prior songs . Based on the direction of the differences between the observed and expected values, the song appears to be more complex.

True/False Questions

1. Nonparametric tests do not rely on means, but they do rely on variances.

2. You should consider using a nonparametric test with highly skewed data.

3. Chi-square tests assess significant differences in frequencies of only one category.

4. Chi-square significance is not determined by probability.

5. The logic for the nonparametric tests is different from the logic for the parametric tests.

6. If an IV has 10 categories, the *df* for the chi-square is 10.

7. There are no two-tailed chi-square tests.

8. The chi-square distribution is very similar to the *t* distribution.

9. The chi-square test is not sensitive to small frequencies.

10. Expected frequencies can be difficult to find in real data.

<u>True/False Answers</u>

1. False	5. False	9. False
2. True	6. False	10. True
3. False	7. True	
4. False	8. False	

Short-Answer Questions

1. In which situations are parametric tests not appropriate?

2. What is a nonparametric test? What are some of the questions that can be addressed using a nonparametric test?

3. What is the difference in what is compared in an ANOVA, in a *t* test, and in a chi-square test?

4. Explain the logic behind the formula of the chi-square test.

5. Why can it be difficult to discover the expected frequencies?

6. Why are chi-square tests always one-tailed?

<u>Answers</u>

1. Parametric tests are not appropriate when using data that are skewed, when the sample size is very small, and finally when the data are not on an interval scale.

2. A nonparametric test does not require the use of means and standard deviations. These tests enable questions to be addressed concerning frequencies and categories.

3. An ANOVA compares the means and variances of multiple categorical groups. A *t* test compares the means and variances of two categorical groups. A nonparametric test compares the observed frequencies in categorical groups to expected frequencies.

4. The logic behind the formula is identical to that of previous hypothesis tests. The deviation between the expected frequency (what you expected) and the observed frequency (what you got) is compared to what was expected (expected frequency rather than standard error in this situation).

5. Expected frequencies are based on information that is obtained prior to conducting the study. These expected frequencies can be difficult to find in research situations because no one has collected data on the categories of interest and the data are difficult to observe in the real world. Thus, there is not a readily available source for the data, and the researcher cannot easily create a data source.

6. Chi-square tests are always one-tailed because the deviation between the observed and expected frequencies is squared. This will always result in a positive value, so the chi-square is always positive.

Multiple-Choice Questions

1. What scale of measurement is most commonly used for the independent variables in a chi-square test?

    a. Nominal   c. Ordinal

    b. Interval   d. Ratio

2. The shape of the chi-square distribution for significant difference is

    a. Positively skewed   c. Symmetrical

    b. Negative skewed   d. Bi-modal

3. What is the smallest possible chi-square value?

    a. 1   c. 0

    b. 0.5   d. −1

4. Which of these can be a parameter?

    a. Mean   c. Variance

    b. Standard deviation   d. All of the above

5. You expected 10 scores in category A and 30 scores in category B, but obtained 7 scores in category A and 33 scores in category B. What is the chi-square for this situation?

    a. $\chi^2 = 0.20$ c. $\chi^2 = 0.30$

    b. $\chi^2 = 1.20$ d. $\chi^2 = 1.30$

6. What is the chi-square for the following table?

A		B	
Observed	Expected	Observed	Expected
112	120	153	145

    a. 0.67   c. 1.21

    b. 0.97   d. 2.69

7. A department store has two entrances, a main entrance and a side entrance. The owner suspects that 75% of his customers enter via the main entrance. At the end of the day, he determines that of his 423 customers, 189 entered through the side entrance. Was the owner correct in his assumption?

    a. Yes he was correct, $\chi^2(1, n = 43) = 86.72, p < .05$

    b. No, he was incorrect, $\chi^2(1, n = 43) = 86.72, p < .05$

    c. Yes he was correct, $\chi^2(1, n = 43) = 86.72, p > .05$

    d. No, he was incorrect, $\chi^2(1, n = 43) = 86.72, p > .05$

8. Larger expected values:

   a. Are related to a greater chance of finding a significant chi-square

   b. Are related to a reduced chance of finding a significant chi-square

   c. Are unrelated to a significant chi-square because it depends on the difference between the observed and expected values

   d. Are usually due to a small standard deviation

9. A college admissions person suspects that having a pet is related to becoming a veterinarian. To test this hypothesis, he polls the past 100 graduates who have become veterinarians and asks them if they own a pet. If this hypothesis were not true, an equal number of veterinarians would have pets as in the general public, which is about 50%. If the study finds that 78 veterinarians have pets, what is the chi-square for this example?

   a. 3.214

   b. 31.36

   c. 24.32

   d. 45.36

10. A colleague of the person in question 9 suggests that this question would be better assessed by comparing these results to graduates who are non-veterinarians. The admissions officer agrees and collects information from 100 non-veterinarians. In this sample of non-veterinarian graduates, 64 have pets and 36 do not own pets. With this new test, the admissions officer expects that 78 of the veterinarians will have pets and 22 will not (the information from the previous question). The admissions officer suggests that if being a veterinarian does not influence pet ownership, then the same number of non-veterinarians should own pets as veterinarians. What would be the chi-square for this new test?

Pet Owner		Non-Pet Owner	
Observed	Expected	Observed	Expected
64	78	36	22

a. 11.42   c. 0.25

b. 23.1   d. 1

<u>Multiple-Choice Answers</u>

1. A	6. B
2. A	7. B
3. C	8. C
4. D	9. B
5. B	10. A

Module Quiz

1. As the owner of a clothing store, you must purchase new shirts in multiple colors: blue, red, pink, and yellow. You purchase an equal amount of each (25%) of each color. After ordering 1000 shirts, you notice that the different colors have sold at different rates. The blue has sold 186, the red has sold 400, the pink has sold 289, and the yellow has sold 125. Where you correct in ordering equal numbers of each color shirt?

    a. What are the hypotheses?

    b. Create a table that shows your comparison.

    c. Calculate χ^2.

    d. Interpret your results at a .01 error level.

2. The head chef is preparing a number of dishes for a banquet of 1000 people. He arrives with enough ingredients to prepare for the following food preferences: 10% vegetarians, 35% chicken, 25% beef, 15% pork, and 15% fish. He has purchased additional ingredients in case the preferences change slightly from expectation, but he will need a lot more if his observation is very different from what he has been told to expect. Upon arriving, he finds out that the requests are as follows: 97 vegetarian, 142 chicken, 320 beef, 210 pork, and 231 fish. Should the chef send out his assistant to purchase additional ingredients?

a. What are the hypotheses?

b. Create a table that shows your comparison.

c. Calculate χ^2.

d. Interpret your results at a .01 error level.

3. A farmer is somewhat concerned about his apple crop this year. He covered his orchard with 30% Gala apples, 20% Red Delicious, 30% McIntosh, and 20% Empire. However, he harvested a total of 1050 apples: 312 Gala, 126 Red Delicious, 433 McIntosh, and 179 Empire. Should he be surprised by his yield?

a. What are the hypotheses?

b. Create a table that shows your comparison.

c. Calculate χ^2.

d. Interpret your results at a .01 error level.

4. What is the chi-square for the following table

A		B		C	
Observed	Expected	Observed	Expected	Observed	Expected
320	335	321	289	214	231

5. Which is the correct presentation of the results from the following table, in APA format?

A		B		C	
Observed	Expected	Observed	Expected	Observed	Expected
10	7	12	15	8	8

Quiz Answers

1a. Null: The shirts sold evenly across all colors.

Alternative: The shirts did not sell evenly across all colors.

1b.

Blue		Red		Pink		Yellow	
Observed	Expected	Observed	Expected	Observed	Expected	Observed	Expected
186	250	400	250	289	250	125	250

1c. $\chi^2 = 174.97$.

1d. Critical $\chi^2 = 11.34$. Reject the null. The shirts did not sell evenly across all of the colors.

2a. Null: There is no difference between the actual requests and expected requests.

Alternative: There is a difference between the actual requests and expected requests.

2b.

Vegetarian		Chicken		Beef		Pork		Fish	
Obs.	Exp.	Obs.	Exp.	Obs.	Exp.	Obs.	Exp.	Obs.	Exp.
97	100	142	350	320	250	210	150	231	150

2c. $\chi^2 = 211.04$.

2d. Critical $\chi^2 = 15.09$. Reject the null. The requests at the banquet differ significantly from what the chef expected. He should send out the assistant for more ingredients.

3a. Null: The amounts of apples that were harvested were in line with how the farmer planted.

Alternative: The amounts of apples that were harvested were not in line with how the farmer planted.

3b.

Gala		Red Delicious		McIntosh		Empire	
Observed	Expected	Observed	Expected	Observed	Expected	Observed	Expected
312	315	126	210	433	315	179	210

3c. $\chi^2 = 82.41$.

3d. Critical $\chi^2 = 11.34$. Reject the null. The farmer's harvest was significantly different from his expectations.

4. 5.47.

5. $\chi^2(2, n = 29) = 1.89, p > .05$.

Module 31

Learning Objectives

- Understand the similar logic underlying various test statistics

- Know the assumptions underlying a χ^2 test of independence

- Find the expected frequencies from row and column totals

- Determine degrees of freedom

- Calculate a two-variable χ^2

- Use a table to interpret χ^2

- Report results in APA format

Module Summary

- Similar to the manner in which an ANOVA can assess group differences among two independent variables, a chi-square test can include a second IV. When a chi-square test contains two independent variables, it is known as a ***test of independence***. The purpose of this test is to determine if the first IV is related to or independent from the second IV.

- A test of independence requires the expected frequency of each cell to be at least 5, regardless of the observed frequency. This is because small expected frequencies have similar difficulties to that of small sample sizes in parametric tests; the chances of making an error are increased. Although there are correction factors that can be applied to a chi-square test in which an expected frequency is less than 5, you should try either to conceptualize your groups in a different manner or to select a different analysis.

- In doing a test of independence, your variables must be independent. Variables that are not independent of one another will influence your results. Imagine trying to determine if enrollment in a specific section of a psychology course is related to the dorms in which students live. You should expect that this decision will occur at random; however, students in the same dorm may provide each other information about the different professors that would unfairly influence the sections in which the students enroll.

- The formula for the test of independence is the same as that for a one-variable chi-square test. The only difference is that the total number of cells is now the product of the number of categories of each of the two factors.

- In conducting a test of independence, first find the expected frequencies. These can be obtained from information provided to you. Alternatively, you can obtain expected frequencies using a formula that multiplies the total frequency for a column by the total frequency for the row and then divides by N. After obtaining the expected frequencies, the chi-square statistic is then calculated using the same formula as that for the one-variable chi-square. This obtained result is then compared to a value obtained from the table in Appendix F. The df used for this test is (number of columns $-$ 1) $\times$ (number of rows $-$ 1). If the value obtained for chi-square exceeds the value obtained from the table, then the frequency proportions are not considered to be the same, and the variables are said to interact.

- Results for tests of independence are expressed in APA format as the following: $\chi^2(df, n) =$ chi-square value, p value; for example, $\chi^2(2, n = 500) = 56, p < .01$.

Computational Exercises

1. An amusement park developer wonders if male and female patrons differ in their preference for the type of ride. The choices are thrill rides, simulators, and ride-through shows. The developer surveys 100 males and 80 females. Of the 100 males, 50 prefer thrill rides, 38 prefer simulators, and 12 prefer ride-through shows. Of the 80 females, 32 prefer thrill rides, 30 prefer simulators, and 18 prefer ride-through shows.

a. What is the null hypothesis?

b. Create a table that shows your comparison.

c. Calculate χ^2.

d. Interpret your results at the .05 error level.

2. A doctor notices that many of her patients are teachers with the flu. The doctor wonders if being a teacher is related to having the flu. In the past week, the doctor saw a total of 70 patients, 14 of whom were teachers. Also, of the teachers, all 14 had the flu, whereas only 13 of the other patients had the flu. Does it appear that being a teacher is related to having the flu?

a. What is the null hypothesis?

b. Create a table that shows your comparison.

c. Calculate χ^2.

d. Interpret your results at the .05 error level.

3. A cookbook company is curious about which is the better food to be paired with peanut butter: jelly or bananas. Its researchers ask a group of 40 people to rate which they prefer. The researchers notice that there appear to be two age demographics among those whom they asked: half of the sample is between the ages of 10 and 15, and the other half is between the ages of 30 and 35. The company expected there to be an equal preference across all groups. If 15 of the younger group prefer jelly and 12 of the older group prefer bananas, does there appear to be a relation between age and preference?

a. What is the null hypothesis?

b. Create a table that shows your comparison.

c. Calculate χ^2.

d. Interpret your results at the .05 error level.

4. A researcher wants to determine if there is a gender difference for beer and wine preference. The researcher hypothesizes there will be no difference between the genders. A group of 50 men and 50 women were asked to rank their preferred drink; 45 men ranked wine as their preferred drink and 35 women ranked beer as their preferred drink. Does gender appear to be related to alcoholic beverage preference?

   a. What is the null hypothesis?

   b. Create a table that shows your comparison.

   c. Calculate χ^2.

   d. Interpret your results at the .05 error level.

5. A professor records the type of shoes worn by male and female students in his summer courses. Of the 68 males, 32 wear sneakers and 36 wear sandals. Of the 95 females, 22 wear sneakers and 73 wear sandals.

   a. What is the null hypothesis?

   b. Create a table that shows your comparison.

   c. Calculate χ^2.

   d. Interpret your results at the .05 error level.

Computational Answers

1a. There is no difference between the sexes in the type of amusement park ride they prefer.

1b.

	Thrill	Simulators	Ride-Through
Male	50 (45.6)	38 (37.8)	12 (16.7)
Female	32 (36.4)	30 (30.2)	18 (13.3)

1c. $\chi^2 = 3.919$.

1d. Critical $\chi^2 = 5.99$. Retain the null. Males and females do not differ in their amusement park ride preferences.

2a. Teacher status is independent of flu status. Or, there is no difference in flu status between those who are teachers and those who are not teachers.

2b.

	Teacher	Not Teacher
Flu	14 (5.4)	13 (21.6)
No Flu	0 (8.6)	43 (34.4)

2c. $\chi^2 = 27.87$.

2d. Critical $\chi^2 = 3.84$. Reject the null. Having the flu does appear to be related to being a teacher.

3a. Age is independent of preference for peanut butter addition. Or, there is no difference in addition preference due to age.

3b.

	Jelly	Bananas
Age 10-15	15 (11.5)	5 (8.5)
Age 30-35	8 (11.5)	12 (8.5)

3c. $\chi^2 = 5.01$.

3d. Critical $\chi^2 = 3.84$. Retain the null. Age is unrelated to preference for peanut butter addition.

4a. Gender is independent of drink preference. Or, there is no difference in drink preference between males and females.

4b.

	Beer	Wine
Male	5 (20)	45 (30)
Female	35 (20)	15 (30)

4c. $\chi^2 = 37.50$.

4d. Critical $\chi^2 = 7.81$. Reject the null. Alcoholic drink preference differs across males and females.

5a. There is no difference between the sexes in the type of shoe they wear to summer classes.

5b.

	Sneakers	Sandals
Male	32 (22.5)	36 (45.5)
Female	22 (31.5)	73 (63.5)

5c. $\chi^2 = 10.22$.

5d. Critical $\chi^2 = 3.84$. Reject the null. Males and females differ in the type of shoe worn to summer classes.

True/False Questions

1. The purpose of a test of independence is to determine if two variables are related.

2. A goodness-of-fit test determines whether two groups are independent.

3. For a test of independence that has 4 categories on one IV and 3 categories on the second IV, the df is 6.

4. In a test of independence, the two IVs must be dependent.

5. A test of independence may include only one variable.

6. In a test of independence, degrees of freedom increase as the sample size increases.

7. The formula for degrees of freedom in a test of independence is $(C - 1) + (R + 1)$.

8. A test of independence compares frequencies across two variables.

True/False Answers

1. True
2. False
3. True

4. False
5. False
6. False

7. False
8. True

Short-Answer Questions

1. What piece of a factorial ANOVA is a test of independence most similar to? Why is this?

2. What additional criteria are required for a test of independence?

3. Why must the expected cell frequency in a test of independence be at least 5? How can this be corrected should an expected frequency be less than 5?

4. What is the difference between a test of independence and a test for goodness of fit?

5. A researcher is interested in determining if there is a gender difference in whether or not people play video games. The researcher also hypothesizes that there may be a role of age, such that adolescents play more than adults. Is this a test of independence or of goodness of fit?

6. A principal wants to determine if her school's teachers are inflating grades by giving too many grades of A versus lower grades. Would the test she conducts be a goodness-of-fit test or a test for independence?

7. A study of 100 participants recently obtained $\chi^2 = 34$ with 3 categories of the first independent variable and 2 categories of the second independent variable. At a Type 1 error rate of .01, how should this result appear in APA format?

Answers

1. It is most similar to the interaction portion of the ANOVA. The interaction states that the influence of the IV on the DV is dependent upon a second IV. A test of independence states that the frequency of one group is dependent upon the frequency of another variable.

2. A test of independence must have an expected value of greater than 5 in each cell, and the IVs must be independent from one another.

3. This is because small cell values can result in an inflation of the error term when calculating the chi-square. This is similar to how small sample sizes can affect parametric tests. If it is not possible to collect more data, then the problem can be corrected by redefining or reconceptualizing categories of the IVs, or you could select a different analysis.

315

4. A test of independence seeks to determine if placement in one category is dependent upon placement in another category. A goodness-of-fit test determines if the chances of falling in various categories are proportional to known or specified proportions.

5. This is a test of independence. There are two variables, age and gender.

6. This is a test for goodness of fit, because there is one variable, grade.

7. $\chi^2(2, n = 100) = 34, p < .01$.

Multiple-Choice Questions

1. A researcher is interested in determining if different subtypes of schizophrenia are related to the number of hallucinations a person has per week. The researcher observes four subtypes of schizophrenics for 10 weeks. What analysis should the researcher use?

   a. Goodness of fit

   b. Test of independence

   c. ANOVA

   d. t test

2. A researcher is interested in determining if there is an interaction between drinking coffee (any amount) and job type. What test would be most appropriate to assess this question?

   a. Goodness of fit

   b. Test of independence

   c. ANOVA

   d. t test

3. Below are the results from the study posed in the previous question. What is the chi-square?

	Doctor		Lawyer		Professor	
	Observed	Expected	Observed	Expected	Observed	Expected
Coffee Drinker	8	10	7	10	15	10
Non-Coffee Drinker	12	10	13	10	5	10

   a. 3.2

   b. 4.3

   c. 7.6

   d. 4.1

4. How would you interpret the results of the previous question?

   a. Being a coffee drinker is related to one's occupation

b. Overall, doctors are less likely to be coffee drinkers than are lawyers

c. Professors are the least likely to be coffee drinkers

d. Being a coffee drinker causes one to be a professor

5. For the table in question 4, how many degrees of freedom were used?

a. 6

c. 4

b. 5

d. 2

6. What is the chi-square for the following table?

	A		B	
	Observed	Expected	Observed	Expected
C	5	6	10	7
D	7	5	2	3

a. 4.51

c. 1.90

b. 5.41

d. 3.1

7. Using the table in question 6, how many degrees of freedom are involved in finding the critical chi-square?

a. 1

c. 3

b. 5

d. 2

Multiple-Choice Answers

1. C

5. D

2. B

6. C

3. C

7. A

4. A

Module Quiz

1. What is the minimum cell expected frequency permitted for any chi-square test?

2. A researcher wants to determine whether the age in which language is first present in a group of children is typical. What is needed to conduct a goodness-of-fit study to assess this hypothesis?

3. An amusement park's owners are interested in determining the size of the families that are coming to the attractions. They count the frequencies of the families depending on how many children the families have (0, 1, 2, 3, 4, 5, or more than 5). Select another categorical variable that the amusement park could consider in the analysis that may or may not be related.

4. A research group is interested in determining if education is related to political affiliation. Its members hypothesize that there will be no difference among education status and political party membership. They develop three categories of education: those with a graduate degree, those with a college degree, and those without any degree. They ask 150 individuals (50 in each group) to list their political affiliation (Democrat or Republican). They find that 30 of those with a graduate degree are Democrats, 23 of those with a college degree are Democrats, and 19 of those without a college degree are Democrats. Does it appear that education is related to political affiliation?

    a. What is the null hypothesis?

    b. Create a table that shows your comparison.

    c. Calculate χ^2.

    d. Interpret your results at the .05 error level.

5. A campground owner wonders if family composition is related to the type of shelter the campers use: tent, pop-up, trailer, or motor home. Family composition is classified as adults only, or adults with children. Here are the campground's reservations:

	Tent	Pop-Up	Trailer	Motor Home
Adults Only	27	12	44	32
Adults with Children	34	78	63	35

a. What is the null hypothesis?

b. Create a table that shows your comparison.

c. Calculate χ^2.

d. Interpret your results at the .01 error level.

Quiz Answers

1. 5.

2. Known percentages of children who first use languages at various ages.

3. Answers will vary. Examples might be income categories or ethnic categories.

4a. Political affiliation is unrelated to education status.

4b.

	Grad Degree	College Degree	No Degree
Democrat	30 (24)	23 (24)	19 (24)
Republican	20 (26)	27 (26)	31 (26)

4c. $\chi^2 = 4.96$.

4d. Critical $\chi^2 = 5.99$. Retain the null. Education status is independent of political affiliation.

5a. There is no difference in the type of campground shelter used by families with different types of composition.

5b.

	Tent	Pop-Up	Trailer	Motor Home
Adults Only	27 (21.6)	12 (31.8)	44 (37.9)	32 (23.7)
Adults with Children	34 (39.4)	78 (58.2)	63 (69.1)	35 (43.3)

5c. $\chi^2 = 27.27$.

5d. Critical $\chi^2 = 11.34$. Reject the null. There is a significant relationship between the type of campground shelter and family composition.

Module 32

Learning Objectives

- Distinguish between statistical significance and practical or clinical importance

- Understand the influence of sample size on statistical significance

- Know which measure of effect size is appropriate for each test of statistical significance

- Calculate various measures of effect size

- Know the guidelines for interpreting effect size

Module Summary

- All of the statistics that have been covered thus far have enabled us to make dichotomous decisions about hypotheses: The observed difference was large, or it was not large. It is important, however, to ask another question: "Does this difference matter?" This is separate from finding a statistically significant difference. You have many have noticed that very large samples will make almost any difference appear statistically significant. The important question for research that statistics hopes to answer is "Does this treatment work? Is it effective?" It should be clear that a hypothesis test does not fully answer this question. Measures that do attempt to answer the question of "Is this an effective treatment?" are called measures of *effect size*.

- Measures of effect size are largely subjective. They do not provide an outcome that states "This is effective" or "This is not effective." It is up to the researcher to determine what constitutes an effective treatment. Measures of effect size simply provide a good starting point.

- One measure of effect size is called *Cohen's* **d**, which rescales the difference between the means of two treatment groups to standard deviation units. The formula for Cohen's *d* is:

$$d = \frac{M_1 - M_2}{s_{DV}}$$

- At first glance, the formula for Cohen's *d* appears very similar to that of an independent sample *t* test. The formula for Cohen's *d*, however, uses a standard deviation in the denominator rather than a standard error. By using a standard deviation from the study's data, the Cohen's *d* refers to this particular study.

- There are not set cutoff points for interpreting a Cohen's *d*. The creator of the statistic suggested that 0.2 standard deviations indicated a small effect, 0.5-0.8 a medium effect, and anything greater than 0.8 a large effect.

- A second measure of effect size is **r**. The statistic *r* measures the extent to which two variables are related. In terms of effect size, *r* represents the association between the IV and the DV. This association ranges from 0 to 1, with 0 indicating no relationship. The formula for effect size *r* is:

$$r = \sqrt{\frac{t^2}{t^2 + df}}$$

- A third measure of effect size is **eta** (η). Similarly to *r*, this effect size ranges from 0 to 1. This formula asks you to determine the proportion of between treatment variability to total variability. In other words, it asks the question "How much of the total variation is attributed to differences among the groups?" The formula is as follows:

$$\eta = \sqrt{\frac{SS_{bet}}{SS_{tot}}}$$

- For a chi-square test, the measures of effect size vary depending on which test you have conducted, a goodness-of-fit test or a test of independence. For a goodness-of-fit test, effect size is calculated with the following formula:

321

$$\text{Effect Size } r = \sqrt{\frac{\chi^2}{(N)(C-1)}}$$

- For a test of independence with two categories per independent variable (2 × 2), the effect measure is called **phi** (ϕ). Phi is interpreted in a similar manner to that of r. This formula is as follows:

$$\phi = \sqrt{\frac{\chi^2}{N}}$$

- When a test of independence has more than 2 categories per IV, the above formula becomes somewhat more complicated and uses a new statistic. This statistic is called a **Cramer's V**. This formula includes phi, which is then amended for the number of categories. The formula is as follows:

$$V = \sqrt{\frac{\phi^2}{(\text{Variable with the fewest of categories}) - 1}}$$

Computational Exercises

1. A study was comparing a new drug therapy for pain management by comparing those on the new medication to those on no medication. The mean pain rating for those on the new drug was a rating of $M = 67$. The mean pain rating for those not on the drug was $M = 78$. The results for this test were significant at the .05 level, but the researcher is uncertain if this means the treatment was effective. If the overall standard deviation was 7, what is the effect size for this study, and how would you classify it?

2. A pharmaceutical company is interested in developing a new drug for depression. Its researchers compare the depression symptoms of those who are not on the medication to those who are on the depression medication. They obtain a difference in means of 3.1 with

an overall standard deviation of 3.7. What was the effect size for this study, and how would you classify it?

3. A scientist is interested in looking at how the brain reacts to stress. He compares the activity of the pre-frontal cortex in a sample of 40 individuals who are under stress to that of 40 individuals who are not under stress. For the stress group, he obtains a mean of 25 points of activity. For the nonstress group, he obtains a mean of 17 points of activity. Although this difference is significant, he is hesitant to report his finding until he has obtained an effect size. If the overall standard deviation was 15.1, how big of an effect does being stressed have on brain activity?

4. A study is interested in the effects of a new therapy for Generalized Anxiety Disorder (GAD). Unfortunately, the study is able to recruit only a small number of participants ($N = 10$). The researchers compare the anxiety of those who received the new therapy ($M = 12$) to those who did not ($M = 18$) and obtain a nonsignificant result at the .05 level. The overall standard deviation for all participants was 5.2. Should the researchers be discouraged with their results?

5. Despite having a good effect size, the researchers in question 4 receive some criticism for their study from the larger psychological community. If they obtained a t value of 6.5, what other measure of effect size could they use to bolster their argument, and what would its value be?

6. A store is trying to increase its amount of customer traffic. To increase traffic, the store runs a special sales promotion in which the first 1000 customers will receive a gift. The store then compares the number of people who entered their store (which was 1240) to the traffic levels

of its 3 top competitors and obtains a $\chi^2 = 7.89$. What is the effect size of this promotion? Did this promotion seem to affect the number of people entering the store?

7. The following week, one of the competitors (Store 2) from the store in question 6 (Store 1) received only 700 customers during the promotion weekend. In order to increase customers, the store ran its own promotion the following weekend and obtained 1231 customers. On this weekend, the store from question 6 received only 800 customers. An analyst decides to determine if the promotion and store interacted to have a significant increase in customers and obtained a $\chi^2 = 239.01$. What was the effect size for this 2 (store) × 2 (promotion) chi-square? Use information from question 6 to calculate N. How does this compare to the effect size found in question 6?

Computational Answers

1. Cohen's $d = 1.57$. Large effect.

2. Cohen's $d = 0.84$. Large effect.

3. Cohen's $d = 0.53$. Medium effect.

4. Cohen's $d = 1.15$. Large effect. The researchers should be pleased because of the large effect size, even though the difference was not statistically significant.

5. $r = 0.92$. This also is a large effect.

6. The effect size $\square = 0.08$. The advertisement appeared to have a minimal effect on the amount of traffic as compared to the competitors. The previous significance was likely due merely to large sample size.

7.

$$\phi = \sqrt{\frac{12.54}{3971}} = 0.06; V = \sqrt{\frac{0.06^2}{2-1}} = 0.06$$

N is the total $N = 1240 + 700 + 1231 + 800 = 3971$.

The effect size is smaller than the effect size found in question 6.

True/False Questions

1. A t test will tell you whether or not a new therapy is effective by showing that the therapy group is different from a placebo group.

2. Measures of effect size provide definitive measurements of the size of the effect.

3. A Cohen's d differs from a t test in that it uses a standard deviation in the denominator instead of a standard error.

4. A Cohen's d of 0.95 is a medium effect.

5. An $r = 1$ indicates no effect.

6. The calculation of effect size for a chi-square test depends on which test you used.

7. Cohen's d and r should provide you with the same exact value for effect size.

8. A Cohen's d of 0.43 and an η of 0.43 have the same meaning.

True/False Answers

1. False	4. False	7. False
2. False	5. False	8. False
3. True	6. True	

Short-Answer Questions

1. What is the next question that is asked after obtaining a significant result? Why is this question important?

2. How can a researcher "force" a result to be statistically significant?

3. How does Cohen's d measure effect size?

4. What about Cohen's d differentiates it from a t test? How does this change the interpretation and utility of Cohen's d?

5. How does *r* measure effect size?

6. What is eta? Why does the formula include a square root?

7. What does it mean when a test is statistically significant but has a very small effect size?

8. What are phi and Cramer's *V*? When is it appropriate to use either?

9. What is the scale that eta, phi, Cramer's *V*, and *r* use to measure effect size? What are its limits?

Answers

1. The next question that should be asked is, "Was this difference meaningful? How large of an effect did the IV have on the DV?" This question is important because it provides more insight into the research question. Any statistic can be significant with a large enough sample. Effect size helps us determine if the significant difference was meaningful or superfluous.

2. A researcher can force statistical significance by increasing the sample size.

3. Cohen's *d* rescales the difference between two means into standard deviation units. This enables you to determine the distance between the two hypothesized distributions in terms of standard deviations.

4. The denominator of the formula differentiates the two. Cohen's *d* uses a standard deviation, whereas a *t* test uses the standard error. This change enables Cohen's *d* to be compared across multiple samples and research questions, as opposed to a *t* test, which is unique to that particular sample size.

5. The statistic *r* is a measure of effect size that assesses the extent to which two variables, the IV and DV, are related.

6. Eta is a measure of effect size for F tests, such as ANOVA. The formula includes a square root so that the value returned is in original units, as the values that are used in calculating eta are squared units (or area units). Area units are not easily interpreted.

7. This indicates that although the chance of obtaining this mean difference by chance is low, the IV had a minimal impact on the DV. As a result, the utility of the results of the study may be minimal. The statistical significance mostly likely was associated with the large sample size.

8. They are measures of effect size for chi-square tests. Phi is used for 2×2 tests of independence, whereas Cramer's V is used when one of the IVs has more than 2 categories.

9. The scale that is used by these measures is correlation, ranging from 0 to 1.

Multiple-Choice Questions

1. What is considered a medium effect?

a. 0.5-0.8

c. 0.3-0.2

b. 0.4 and greater

d. It depends on how effect size is measured

2. The difference between two means is 7, and the standard deviation for all the scores in the study is 4. What is Cohen's d?

   a. 3.5

       c. 0.5

   b. 1.75

       d. 7

3. An independent-sample t test yields a t of 4.3 with 16 degrees of freedom. What should the researcher conclude about this result (using alpha of .01)?

   a. The result is statistically significant with a large effect

   b. The result is not statistically significant with a large effect

   c. The result is statistically significant with a small effect

   d. The result is not statistically significant with a small effect

4. What is the effect size of the previous question?

    a. 0.55

    c. 0.73

    b. 0.64

    d. 1.23

5. What can an effect size r never be?

    a. 0.5

    c. 1.32

    b. 0

    d. 0.1

6. The results of an ANOVA provide an $SS_{bet} = 582$ and an SS_{tot} of $= 1839$. What is the size of this effect?

    a. 0.56

    c. 0.89

    b. 0.47

    d. 1.2

7. A 3×2 test of independence had an $N = 150$ and a $\chi^2 = 5.3$. What is the size of this effect?

    a. 0.19

    c. 0.63

    b. 0.04

    d. 0.47

8. With the following information $M_1 = 8.47$ and $M_2 = 10.47$ and an $S = 2.68$, what is the effect size?

    a. 2

    c. 1.23

    b. 0.75

    d. 0.98

9. As effect size _____, the sample size needed to obtain adequate power to detect the effect _____.

    a. Increases, increases

    c. Remains the same, increases

    b. Decreases, increases

    d. Increases, remains the same

328

10. In a study comparing time it takes two types of paint dry to on $N = 31$ walls, a t statistic of

 1.46 is obtained. If the standard deviation for the two groups is 10.4, which of the following

 is true?

 a. t is significant at the .05 level and the effect size is $r = .26$.

 b. t is not significant at the .05 level and the effect size is $r = .26$.

 c. t is significant at the .01 level and the effect size is $r = .07$.

 d. t is not significant at the .01 level and the effect size is $r = .07$.

11. An electric company is trying to make a brighter light bulb. The standard light bulbs have an

 average brightness rating of 5.2, and the new light bulbs have an average brightness rating of

 6.9. The standard error between these two groups of light bulbs is 2.1. If there are a total of

 46 light bulbs in this study, which of the following is true?

 a. t is significant at the .05 level and the effect size is $r = .12$.

 b. t is not significant at the .05 level and the effect size is $r = .12$.

 c. t is significant at the .01 level and the effect size is $r = .2$.

 d. t is not significant at the .01 level and the effect size is $r = .2$.

12. Using the following ANOVA source table, what are the effect size for Factor 1 and the effect

 size for Factor 2?

	SS	df	MS	F
Factor 1	657	4	164.25	3.32
Factor 2	327	3	109	2.20
Interaction	412	12	34.33	0.69
Within	3,214	65	49.45	
Total	4,610	84		

a.   Factor 1 $\eta = 0.38$, Factor 2 $\eta = 0.27$ c.   Factor 1 $\eta = 0.45$, Factor 2 $\eta = 0.32$

b.   Factor 1 $\eta = 0.14$, Factor 2 $\eta = 0.07$ d.   Factor 1 $\eta = 0.20$, Factor 2 $\eta = 0.10$

1. D   7. A

2. B   8. B

3. A   9. B

4. C   10. B

5. C   11. B

6. A   12. A

Module Quiz

1. A new toaster oven's manufacturer advertises that it can toast faster than any other toaster! You are skeptical when you receive this toaster as a gift and decide to compare it to your already working toaster by toasting $N = 20$ (10 each toaster) slices of bread. The new toaster takes an average of $M = 68$ seconds to toast bread, whereas your old toaster takes an average of $M = 73$ seconds to toast bread. Your find the standard error to be 2.1. Is this a significant result at a .05 error level? What is the effect size for this result? Does the new toaster's faster toasting seem to matter?

2. A researcher is interested in reducing the intensity of seizures in those with seizure disorder. A new drug is administered to 15 people with seizure disorder, while another 15 people with seizure disorder are monitored without the drug as a comparison. As rated by a third party, the intensity of seizures of those on the medication has a mean of 17. The intensity of seizures of those not on the medication has a mean of 23. If the standard error is 1.20, what is the significance and effect size of the medication?

3. A team of researchers is interested in assessing the effect of an herbal supplement on calcium in bones. They create three groups: (1) herbal supplement, (2) placebo, and (3) control group. They obtain the following results:

	SS	df	MS	F
Between	687	2	343.50	13.14
Within	784	30	26.13	
Total	1,471	32		

What is the size of the effect of the differing treatments?

4. You are interested in studying the effects of a medication designed to suppress the symptoms of HIV. You compare three medications to a placebo group in order to determine the effect. You obtain the following results:

	SS	df	MS	F
Between	347	3	115.67	5.36
Within	647	30	21.57	
Total	994	33		

What is the size of the effect of various medications?

5. A marketing company is interested in determining the influence of a new advertisement for running shoes. It compares the interest (rated 1-10) of those who have seen the advertisement to ratings of those who have not. The company also believes that gender will affect the decision, such that men will be more influenced by the ad than women. Here are the results.

	SS	df	MS	F
Advertisement	68	1	68	4.65
Gender	75	1	75	5.13
Interaction	120	1	120	8.21
Within	658	45	14.62	
Total	921	48		

What is the size of the interaction effect of gender and the advertisement on interest in the shoe?

6. In the previous module, you were asked to determine if there was a relation between gender and preference for different alcoholic beverage (beer or wine). The $\chi^2 = 37.5$, and there was a total of 100 participants. What was the effect size of this analysis?

Quiz Answers

1. $t(18) = 2.38$, $p < .05$. $r = 0.49$. The effect size is close to a medium effect, but a time difference of 5 seconds may not be important unless you are in a big rush!

2. $t(28) = 5$, $p < .01$. $r = 0.22$, small effect. It appears that the medication is not effective, although it is statistically significant.

3. $\eta = 0.68$.

4. $\eta = 0.59$.

5. $\eta = 0.36$.

6. $\square = 0.61$.

Module 33

Learning Objectives

- Know the mathematical relationship between Type 2 error and power

- Understand the effect that various factors have on power

- Understand the interrelationship between statistical significance, effect size, and power

- Use graphs or tables to determine power, given known values of other variables

- Use graphs or tables to determine the effect size, given known values of other variables

- Use the graphs or tables to determine necessary the sample size for desired power, given known values of other variables

Module Summary

- ***Power*** is the ability of a test to correctly detect an effect (correctly reject the null hypothesis). In other words, power is the opposite of a Type 2 error. The power of a study can be calculated by subtracting the probability of making a Type 2 error from 1. Increasing the amount of power in a study generally is desirable for a researcher.

- Power is affected primarily by five things: (1) Type 1 error, (2) the directionality of the test, (3) the size of the effect, (4) error variance, and (5) sample size. As alpha (Type 1 error) increases, the amount of power a study has also increases. This is because moving the alpha level closer to the mean (increasing alpha) allows for a larger region of rejection. With a larger region of rejection, the probability of correctly rejecting the null increases. This is how the directionality of a test will influence power, as a one-tailed test has a larger region of rejection in the single tail than a two-tailed test has in each of its tails. Similarly, as effect size increases, so does power. A larger effect indicates less overlap between the null distribution and the alternative distribution. The less overlap there is, the higher the

333

probability there is of correctly rejecting the null (an increase in power). The reverse is true for error variance. A large amount of error variance will reduce the amount of power in a study. Large samples reduce error variance and so increase the chances of rejecting the null. Thus, large samples increase power, while small samples decrease power. Recall that anything that decreases the size of the denominator in a hypothesis test will increase the chances of finding a significant result, and anything that increases the size of the denominator decreases the chances of funding a significant result.

- Behavioral research is a complex process, and therefore we cannot always expect findings that are as straightforward as an IV being the sole cause of change in a DV. In considering this, it is important to note that as you increase power, the chances of making a Type 1 error also increase. This presents us with a trade-off between power, which is desirable, and low Type 1 error, which is also desirable.

Computational Exercises

1. You are interested in studying the effects of a medication designed to suppress the symptoms of HIV, using a sample of $N = 30$ at an alpha level of .05. You compare three medications to a placebo group in order to determine the effect. You obtain the following results:

	SS	df	MS	F
Between	347	3	115.67	5.36
Within	647	30	21.57	
Total	994	33		

Following these successful results, you replicate the study with a sample size of $N = 60$. Also, the alpha level was decreased to .01. How was power affected by these changes?

2. Using Table 33.2 in the text, if you are conducting an ANOVA with 5 groups, want a power of .8, and expect an effect size between 0.4 and 0.5, how large should your sample be per group?

334

3. Using Table 33.2 in the text, if you are conducting an ANOVA with 4 groups, want a power of 0.9, and expect an effect size between medium and large, how large should your sample be per group?

4. Using Table 33.1 in the text, if you are conducting an independent-sample t test with an effect size of 0.20 (Cohen's d) and have a sample of nearly 400 per group, what is your anticipated power?

5. Using Table 33.3 in the text, if you conduct a chi-square test of independence with 4 df, power of 0.9, and a sample of 388, what is the minimal effect size you will be able to detect?

Computational Answers

1. Power was increased by adding more participants, but it was decreased by increasing the alpha level.

2. The sample should have 11 participants per group.

3. The sample should have 40 participants per group.

4. The power is 0.80.

5. You will be able to detect a minimum of 0.20 effect size.

True/False Questions

1. Power is $1 - \alpha$.

2. Decreasing the sample size will reduce a study's power.

3. A large treatment effect will increase power.

4. Increases in power are always preferred, with a power level of 100% being ideal.

5. A large enough sample will make even the smallest difference between two means appear significant.

1. False   3. True   5. True

2. True   4. False

Short-Answer Questions

1. What is power in a research? Why is it desirable?
2. What are the factors that affect power?
3. Explain how a large effect size and a small error variance work in conjunction to increase power.
4. Explain how increasing the alpha level and changing the directionality of a test from two tailed to one tailed have the same effect on power.
5. Explain the relationship between sample size and power. How does changing sample size change power? How does the size of an effect impact the size of the sample needed to maintain a certain level of power?

Answers

1. Power is a test's ability to correctly reject the null. It is desirable because it improves the accuracy of the test.

2. Five factors affect power: (1) sample size, (2) error variance, (3) directionality of the test, (4) effect size, and (5) Type 1 error level.

3. A large effect size indicates that the population means are very far apart. A small error variance indicates that there is less variability within each group. Combined, these indicate that the scores in each population are clustered around the mean and that the means are far apart, which would result in a situation that would greatly increase the chance of rejecting the null.

4. Changing the directionality of a test changes the region of rejection in one tail. Thus, these principles are essentially the same, as using a one-tailed test instead of a two-tailed test will increase the region of rejection in one specific tail, increasing alpha and power.

5. Increasing sample size will increase power. This is because larger samples better represent a population. If there is a true difference between two populations, then larger samples will

more accurately represent these populations, which will be reflected in a smaller error variance and therefore a higher chance of correctly rejecting the null. As effect size increases, the need for a large sample decreases.

Multiple-Choice Questions

1. Which of the following does not affect power?

    a. Effect size

    b. α

    c. Sample size

    d. The number of independent variables

2. The chance of making a Type 2 error is 0.31. What is power?

    a. 0.31

    b. 0.69

    c. 0.05

    d. 0.80

3. Which element of power does the researcher have the most control over?

    a. Effect size

    b. α

    c. Sample size

    d. Error variance

4. As sample size _____, power _____.

    a. Increases, decreases

    b. Decreases, decreases

    c. Increases, increases

    d. Remains the same, decreases

5. As Type 1 error rate _____, power _____ and the chances of making a Type 2 error _____.

    a. Increases, increases, increases

    b. Decreases, decreases, increases

    c. Increases, decreases, increases

    d. Increases, increases, decreases

Multiple-Choice Answers

1. D

2. B

3. C

4. C

5. D

Module Quiz

1. What is the pitfall of increasing power?

2. How do small error variances and large sample sizes affect power?

3. A researcher is testing a new treatment for PTSD but discovers that his samples have a tremendous amount of variability. What should the researcher worry about doing in this situation, and how could he rectify the problem?

4. How is the directionality of a test related to power?

5. How would increasing the numerator of a hypothesis test affect effect size?

Quiz Answers

1. The main issue with increasing power is that it can lead you to making a Type 1 error. Also, it is difficult to state that the result obtained from your study will hold true in all situations. As a result, the extent to which you can correctly reject a null in one situation may be very different in another situation.

2. They decrease the denominators of hypothesis tests, which increases power.

3. The researcher should be concerned about low power, or falsely failing to reject the null hypothesis of no effect. He could rectify this by including more people in the study.

4. The directionality of a test (changing from a two-tailed to a onr-tailed test) increases power (as long as the effect is in the correct tail!) because it increases the size of the rejection region in that one tail.

5. Increasing the numerator of a hypothesis test means that there is a larger difference between "what you got" and "what you expected." If the denominator remains constant, then increasing the numerator would result in a larger effect size.

Module 34

Learning Objectives

- Distinguish between correlational and experimental studies

- Create a scatterplot

- Estimate the strength and direction of the relationship in a set of data based on its scatterplot

- Understand the effect of outliers on the strength and direction of a correlation

Module Summary

- Correlational studies differ from those that have been covered thus far. Correlational studies contain a single group and seek only to find relations between variables, rather than finding cause and effect. An example of a relation between variables is a *reliability test*. A teacher could give students two different forms of a test and then see if scores on the two forms were related. If they were, this would indicate that the tests were equivalent; however, if the scores were unrelated, then it may be the case that one form was more difficult than the other or that the two different forms don't assess the same things. Another use of a correlational study is *prediction*, or determining how one person will do on Variable A based on his or her performance on Variable B. Using the previous example, imagine that the teacher wants to determine how students will do on the second test. If a previously conducted correlational test had shown two tests to be correlated, the teacher could use the scores from the first test to predict scores on the second test, because the tests had been shown to be about equally difficult.

- One method of displaying correlational studies is through a scatterplot. Scatterplots are graphs that display individuals' scores on two separate variables. One variable is represented on the X-axis, and the other is represented on the Y-axis. Individuals' scores are graphed by using the scores as coordinates on the graph. The relation between the two variables is called a ***bivariate*** relation.

- Correlational studies indicate the strength of a relationship between two variables. The strength, or magnitude, of the relation is measured on a scale that ranges from –1 to 1. A value of 1 indicates the strongest possible positive relation. A ***positive*** relation means that increases on Variable A are related to increases on Variable B. The scatterplot for a positive relationship resembles a straight line starting at the origin of the graph and radiating to the upper right at a 45 degree angle. The closer the relation is to 1, the more the plot will resemble a straight line. Conversely, a value of –1 indicates the strongest possible negative relation. A ***negative*** relation means that increases on Variable A are related to decreases on Variable B. The scatterplot for a negative relationship resembles a straight line starting at the top of the Y-axis and moving down to the bottom right at a 45 degree angle until it touches the X-axis. Again, the closer the relation is to –1, the more the plot will resemble a straight line. Finally, a score of 0 indicates no relation. This means that any change on Variable A is unrelated to change on Variable B. The scatterplot for a correlation of 0 would appear very much like a round blob. In the social sciences, correlations of 1 and –1 are rarely achieved.

- Correlations explain relations between two linearly related variables. ***Linearity*** means that a 1-unit increase in Variable A is related to a consistent amount of increase in Variable B. When linearly related variables are graphed on a scatterplot, the points resemble a straight line. Variables may also be related in a ***curvilinear*** fashion. This indicates that an increase on Variable A is related to varying levels of change on Variable B. When curvilinear variables are drawn on a scatterplot, the points resemble a curved line.

- Correlations may be heavily influenced by outliers. *Outliers* are extreme scores, or scores that are very different from the rest of the data. An outlier can significantly alter the correlation between two variables.

True/False Questions

1. A scatterplot allows us to estimate the strength and direction of the relationship between two variables.

2. A significant correlation with a large sample definitely means that the variables are related.

3. A positive correlation indicates that as scores on Variable A increase, scores on Variable B decrease.

4. Two uses of a correlation are reliability and prediction.

5. The strength of a correlation refers to the extent to which changes in one variable are related to changes in another variable.

6. Outliers can change the strength of a correlation.

True/False Answers

1. True	3. False	5. True
2. False	4. True	6. True

Short-Answer Questions

1. How does a correlational study differ from a *t* test or ANOVA?

2. What are two uses of correlational studies? How are these two use different?

3. Why are correlations of 1 and –1 rarely found in the social sciences?

4. What does a negative correlation indicate? What does a positive correlation indicate?

5. What is the difference between a linear relation and a curvilinear relation?

Answers

1. A correlational study measures the extent to which two variables are related or vary together. An ANOVA and a *t* test measure the causal relationship between two variables.

2. Correlational studies can be used for prediction and reliability. Prediction is using a person's score on one variable to estimate the score on a second variable. Reliability is determining if measurements on two separate tests agree with one another. Using the results of a correlational study for reliability means that the researcher wants to find out whether two variables are related to each other, whereas using the results for prediction means that the researcher will use the value of one variable, along with the strength and direction of the relationship between it and a second variable, to predict the value of the second variable. The studies also differ in the level of correlation that is acceptable. Predictive studies usually allow a lower correlation than do reliability analyses.

3. A correlation of 1 or –1 implies a perfect linear relationship, meaning that scores on one variable directly correspond to scores on another. In the social sciences, having such a high level of precision is rare, which reduces the strength of the relationship. Social science research measures tend to contain a fair amount of error, which reduces the strength of relationships and makes correlations closer to zero.

4. A negative correlation indicates that increases on one variable are related decreases on the other. A positive correlation indicates that increases on one variable are related to increases on the other.

5. A linear correlation indicates that the relation between the variables follows a straight line, whereas a curvilinear relation indicates that the relation between the variables curves. In a linear relation, a 1-unit increase in Variable A is thought to be related to a constant amount of

increase in Variable B. In a curvilinear relation, a 1-unit increase in Variable A is related to varying amounts of increase in Variable B.

Multiple-Choice Questions

1. Bivariate relationships are relationships involving

    a. One variable

    b. Two variables

    c. Three variables

    d. Four or more variables

2. As a correlation increases in strength, the points on the plot begin to _____.

    a. Become closer together, forming a line

    b. Become spread out, forming a "blob"

    c. Fan out at one end

    d. Demonstrate causation

3. Below are the data representing ratings of attractiveness and the amount of cologne a male was wearing. How would you describe the relationship?

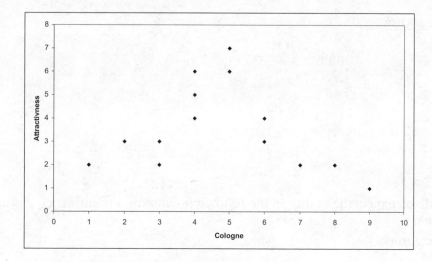

    a. A weak, linear, negative relationship

    b. A strong, linear, positive relationship

    c. A moderate, curvilinear relationship

    d. A weak, linear positive relationship

4. If the actual relationship between two variables is curvilinear but you calculate a linear relationship on the data, which of these will be true?

    a. The calculated value will be lower than that of the actual relationship

    b. The calculated value will be higher than that of the actual relationship

    c. The calculated value will be in the opposite direction (+ or –) from that of the actual relationship

    d. It is not possible to calculate a linear relationship on curvilinear data

5. What type of relationship would you expect between outdoor temperature and the amount of water consumed by people, on average?

    a. A strong positive relationship   c. A strong negative relationship

    b. A weak positive relationship   d. A weak negative relationship

Multiple-Choice Answers

1. B   4. A

2. A   5. A

3. C

Module Quiz

1. How would you expect the outlier in the following scatterplot to influence the strength of the linear association?

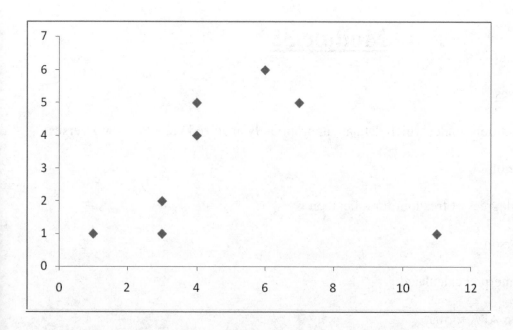

2. How do variables vary together in a positive relationship?

3. What are the limits of the range of a correlation coefficient?

4. How can a correlation be used for prediction?

5. How does a correlation assess reliability?

Quiz Answers

1. The outlier in the distribution would decrease the strength of the linear association.

2. Decreases (increases) on one variable are related to decreases (increases) on another.

3. A correlation can range in value from –1 to 1.

4. Correlations provide a measure of the strength of the relation between two variables, such that if we know the score one variable (X), we can more accurately determine the score on a second variable (Y). The strength of the correlation lets us know how accurately we can predict Y if we know X.

5. A correlation assesses reliability by providing evidence that scores on one form of a variable correspond to scores on another form of the same variable. For example, it might demonstrate that students who score poorly on Form A of an English test also will score poorly on Form B of the English test.

Module 35

Learning Objectives

- Distinguish conditions under which data are appropriately analyzed via a Pearson r versus another correlation

- Determine the degrees of freedom for a Pearson r

- Calculate a Pearson r

- Use a table to interpret calculated r

- Report results in APA format

Module Summary

- The strength and direction of a correlation are measured by a ***correlation coefficient***. As mentioned previously, the strength of a correlation is measured on a scale ranging from -1 to 1, with 0 indicating no correlation between the variables. Multiple types of correlation coefficients exist, but all have a similar function of measuring the relationship between two variables. The correlation coefficient that will be focused on here is a ***Pearson* r**. This is used when the relationship between two variables is linear and the data have been measured on either an interval or a ratio scale. The raw score formula for a Pearson r is:

$$r_{XY} = \frac{N\sum XY - (\sum X)(\sum Y)}{\sqrt{[N(\sum X^2) - (\sum X^2)][N(\sum Y^2) - (\sum Y^2 0]}}$$

- There are multiple formulas for a Pearson r, one that includes z scores, another that includes deviation scores, and finally the one shown above, which includes raw scores. Using the formula that incorporates z scores, it is possible to graph the z scores for each data point on a 4-quadrant graph. This can be helpful in interpreting positive and negative relationships in scatterplots. For a positive relationship, nearly all of the plotted z scores will fall in the

346

second or fourth quadrant, indicating that most of the scores have coordinates that are matched in sign—either both positive or both negative. As scores increase on one variable, they tend to increase on the other. For a negative relationship, nearly all of the plotted z scores will fall in the first or third quadrants, indicating that most of the scores have one positive coordinate and one negative coordinate.

- After a correlation coefficient is calculated, the next question that should be asked is "Are these variables significantly related, or are they unrelated?" This question is similar to the question we asked for inferential statistics: "Is this difference between means large or small?" Because of this similarity in the questions asked, correlation coefficients can be tested in a process similar to what we have done thus far. In the hypothesis test involving a correlation coefficient, we assess the probability that we would find a relationship among the scores. We can expect that if they are not related, the correlation coefficient would be close to 0. We can expect to find some small relationship between the variables due to chance, but overall, the scores on the two variables are unrelated. Alternatively, if they are related, we would expect the relation between the variables—the correlation coefficient—to be very different from 0.

- Similar to the method used for other inferential statistics, the way we determine a significant relationship is by using a table; the appropriate table can be found in Appendix G. The table is used by first finding the degrees of freedom. The df is $N - 2$ (where 2 is the number of variables). This is used to find the critical value for the correlation. If the obtained correlation exceeds the critical value, then the null is rejected and we have a significant relation. Looking at the table, you may also notice that there are multiple sets of numbers per df, for different Type 1 error rates and directionality of the test. Directionality in a correlation study indicates that the relationship between two variables is expected to be in a particular direction—either positive or negative—rather than the researcher hypothesizing that some significant correlation exists, but not hypothesizing the direction of the relationship.

- After obtaining a correlation, it is common to write the results in the following APA format: $r(df)$ = correlation coefficient, p value; for example, $r(8) = .69, p < .05$.

Computational Exercises

1. A college admissions officer is interested in determining if the difficulty of a course is related to how well students do in the course. The officer hypothesizes that these variables will be positively related, because she thinks that a difficult course inspires a student to work harder. Here are the difficulty ratings, on a scale of 0-20, and the corresponding grade received in the class, on a 0-100 scale.

Difficulty	Grade
7	77
9	77
12	81
14	88
17	90

   a. What is the Pearson r?

   b. Interpret this correlation, explaining the strength and direction of the relationship.

2. A dentist is interested in determining if there is a relationship between the length of time (in minutes) that his patients spend brushing their teeth per day and the amount of plaque that has accumulated on their teeth. He obtains the following data from 7 of his patients.

Time Brushing	Plaque
2	8
4	7
5	4
5	6

4	10
4	6
1	2

a. What is the Pearson r?

b. Interpret this correlation, explaining the strength and direction of the relationship.

3. A construction crew is having a competition to determine who can lift the most weight. The crew members believe that the amount a person can lift is strongly related to the proportion of muscle mass that the person has. Here are the data relating how much each person in the crew of 6 lifted and each person's proportion of muscle mass, obtained at the person's most recent physical.

Weight Lifted (lbs.)	Muscle Mass (%)
152	56
176	58
180	66
228	68
263	72
307	84

a. What is the Pearson r?

b. Interpret this correlation, explaining the strength and direction of the relationship.

4. As the editor of a consumer advocacy magazine, you are interested in determining if the cost of a blender is associated with its price. Using the following data on prices and perceived quality, should you advise your readers to purchase blenders that cost more?

Cost	Quality
31	4
31	2
48	13
31	5
15	4
19	8
38	10
40	13

a. What is the Pearson *r*?

b. Interpret this correlation, explaining the strength and direction of the relationship.

5. You are doing a study on eating disorders and want to determine if there is a relationship between the number of calories people with binge eating disorder consume and the levels of hunger they feel during their binges. You measure hunger by means of a self-report rating (on a 1-10 scale). Using the following data, does it appear that hunger and calories consumed are related?

Hunger Rating	Calories Consumed
2	670
7	1120
0	945
4	551
6	1362
2	597
3	1338

a. What is the Pearson *r*?

b. Interpret this correlation, explaining the strength and direction of the relationship.

6. A magazine recently reported that being physically attractive was related to the number of times a person was arrested. Below is the scatterplot of the data. Do you think that this is an accurate report? Why or why not?

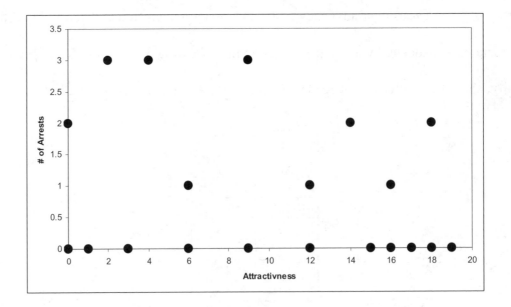

7. A marriage researcher is interested in determining if marital satisfaction is comparable between the partners in a marriage. She obtains data from the husbands and wives in 10 couples to determine the extent to which they are satisfied with their marriage. From the data presented below, does there appear to a relationship between marital satisfaction felt by men and women?

Husband Satisfaction	Wife Satisfaction
2	9
4	34
4	44
7	15
13	19
15	22
25	24
35	11
37	22
43	41

    a. What is the Pearson r?

    b. Interpret this correlation, explaining the strength and direction of the relationship.

8. Depression and anxiety are considered to be highly comorbid (co-occurring) disorders in adults. A researcher is interested in determining if this relationship occurs in children as well. Below are the scores, for 8 children, on a depression and anxiety inventory.

Depression	Anxiety
7	3
8	4
10	6
13	13
13	15
18	24
19	24
19	24

a. What is the Pearson r?

b. Interpret this correlation, explaining the strength and direction of the relationship.

9. Mindfulness is the concept of a person being oriented to the present in his or her thinking. Prior research has shown that those with social anxiety have trouble focusing on what is occurring in social situations, and this, in turn, has been interpreted as a lack of mindfulness. To explore this possible relationship, you collect data on both social anxiety and mindfulness. For each of the two measures, a higher scores indicates that the person exhibits more of the trait.

Social Anxiety	Mindfulness
6	8
7	7
8	7
9	5
10	4
11	4
12	3
14	1
14	1

a. What is the Pearson r?

b. Interpret this correlation, explaining the strength and direction of the relationship.

10. The dean of a university is interested in determining how well teacher competence (as measured by a standardized test) is related to student satisfaction in the classroom. The following data are obtained from 5 teachers.

Competence	Satisfaction
84	10
57	10
85	1
41	10
43	5

a. What is the Pearson r?

b. Interpret this correlation, explaining the strength and direction of the relationship.

Computational Answers

1a. $r = .96$.

1b. There is a strong positive relation between course difficulty and grade, $r(3) = .96, p < .05$.
This indicates that students who do better in courses rate them as being more difficult, or that the more difficult the course, the better students do in that course.

2a. $r = .27$.

2b. There appears to be no relation between the amount of time spent brushing and the amount of plaque accumulated on the patients' teeth, $r(5) = .27, p > .05$. This lack of a relation probably can be attributed to the small size of the sample.

3a. $r = .96$.

3b. There is a strong positive relationship between the proportion of muscle mass and the amount of weight a person can lift, $r(4) = .96, p < 0.05$.

4a. $r = .65$.

4b. There is a moderate positive relationship between the quality of a blender and its cost. This relation was not significant, $r(6) = .65, p > .05$. This may be attributed to the small sample size used.

5a. $r = .44$.

5b. The relationship between hunger and calories consumed is not significant, $r(5) = .44$, $p > .05$.

6. It appears that there is no relationship, because the majority of the participants in the sample had 0 arrests, regardless of their attractiveness. For those who were arrested at least once, the number of arrests is not related to attractiveness; the scatterplot appears to be approximately horizontal.

7a. $r = .07$.

7b. There is no relation between the marital satisfaction levels of husbands and their wives, $r(8) = .07, p > .05$.

8a. $r = .99$.

8b. There is a very strong positive relationship between depression and anxiety in children, $r(6) = .99, p < .05$.

9a. $r = -.99$.

9b. There is a very strong negative relationship between social anxiety and mindfulness, $r(7) = -.99, p < .05$. This indicates that people with more social anxiety exhibit less mindfulness.

10. $r = -0.39$. There is a moderate to low level of negative relationship between teacher competence and student satisfaction, $r(3) = -.39, p > .05$. The relationship is not statistically significant. This may be attributed to the small sample size. Alternatively, the low negative relationship might be the result of an unreliable or poorly suited test of teacher competence or of student satisfaction.

True/False Questions

1. The formula for df for a Pearson r is $N - 1$.

2. A Pearson r measures the strength of the relationship between two linearly related variables.

3. A point biserial correlation measures the relationship between an interval variable and a dichotomous variable.

4. The Pearson *r* is the only method to assess the relationship between two variables.

5. A scatterplot of *z* scores covers four quadrants, whereas a scatterplot of raw scores falls in only one quadrant.

6. The three formulas for calculating a Pearson *r* correlation coefficient (using *z* scores, deviation scores, and raw scores) lead to the same answer.

7. To calculate a Pearson *r* correlation coefficient, both variables must be measured on at least an interval level.

True/False Answers

1. False	4. False	7. True
2. True	5. True	
3. True	6. True	

Short-Answer Questions

1. What is a Pearson *r*?

2. What are the hypotheses for correlation coefficients?

3. What is being tested in a hypothesis test for a correlation?

4. For a positive correlation, in which quadrants should you expect the majority of the points on a scatterplot to fall?

5. You calculate a correlation between two variables for 10 participants and find that the two variables are unrelated. What might explain this result, other than an actual low level of relationship between the two variables?

Answers

1. A Pearson *r* provides a numerical measure of the strength and direction of a linear relationship between two variables.

2. The null hypothesis states that there is no relationship between the two variables, whereas the alternative hypothesis states that there is a relationship between the two variables.

3. The hypothesis test determines whether the variation in one variable is related to the variation in another variable. This is done be determining the probability of observing this covariation by chance and then making a statistical decision.

4. The points should fall in the second and fourth quadrants, with the measures usually having the same sign (both positive or both negative).

5. The seeming lack of a relationship might be the result of the small sample size. To have stronger evidence that there really is no relationship, you could run the study again, using a larger number of participants.

Multiple-Choice Questions

1. The data below are from a dental study assessing the amount of pain felt and the amount given to the patient of a new formula for Novocain. What is the Pearson r?

Pain	Novocain
4	3
1	2
3	1
3	4
2	2

   a. .76   c. 0

   b. .23   d. .35

2. Using a sample size of 15 and a one-tailed test, what is the lowest alpha level at which a Pearson r of .58 is significant?

   a. .05   c. .01

   b. .025   d. .005

3. A researcher is comparing scores on an aptitude test and job performance for 32 individuals. How many degrees of freedom should be used when determining if the relationship is statistically significant?

   a. 32

   c. 30

   b. 31

   d. 34

4. Using the information from the previous question, if the researcher obtained a correlation of .43, which of the following statements should be reported, assuming a two-tailed test?

   a. $r(32) = .43, p > .05$

   c. $r(30) = .43, p < .02$

   b. $r(30) = .43, p < .01$

   d. $r(32) = .43, p < .05$

5. A correlation is reported as $r(17) = .32, p > .05$. How many participants were involved in this study?

   a. 17

   c. 18

   b. 19

   d. 15

6. A Spearman rho measures a relationship between two variables on a(n) _____ scale.

   a. Nominal

   c. Interval

   b. Ordinal

   d. Ratio

Multiple-Choice Answers

1. D

4. C

2. B

5. B

3. C

6. B

Module Quiz

1. An auto company is interested in determining if the weight of a car is related to its the number of miles it gets per gallon of gasoline used. Here are the data from a study measuring the weights of cars (in tons) and their gas mileage (miles per gallon, or MPG).

Weight	MPG
25	38
31	37
31	32
34	32
38	31
40	30
44	27
48	26

a. What is the Pearson r?

b. Interpret this correlation, explaining the strength and direction of the relationship.

2. The Department of Motor Vehicles is interested in determining whether the amount of time people spend waiting in line is related to the amount of frustration they experience at the DMV. Here are the frustration scores of 8 people who waited in line at the DMV, along with the amounts of time they spent waiting.

Time	Frustration
2	2
9	2
11	4
15	8
23	9
28	9
33	10
65	21

a. What is the Pearson r?

b. Interpret this correlation, explaining the strength and direction of the relationship.

3. Using the data from the previous example, do there appear to be any outliers in the data? What effect, if any, might these scores have?

4. A graduate training program in clinical psychology is interested in determining if there is a relationship between a therapist's measured skill and the number of hours he or she has done therapy. Below are the data from 7 student therapists.

Skill	Hours
5	197
7	184
8	218
9	214
12	247
13	232
13	245

a. What is the Pearson r?

b. Interpret this correlation, explaining the strength and direction of the relationship.

5. A graduate admissions program wants to investigate its recruitment procedure to determine if there is a relationship between GRE (Graduate Record Examination) scores and a student's performance in graduate school (as measured by GPA). Below are data on GPAs and GRE scores for 10 graduate students.

GRE	GPA in Grad School
1160	3.25
1180	3.14
1200	3.78
1280	3.90
1290	3.80
1370	4.00
1400	4.00
1520	4.00
1570	3.90
1590	3.70

a. What is the Pearson r?

b. Interpret this correlation, explaining the strength and direction of the relationship.

Quiz Answers

1a $r = -.94$.

1b. There is a strong negative relation, $r(6) = -.94, p < .01$. As the weight of a car increases, the number of miles it can travel per gallon of gasoline decreases.

2a. $r = .98$.

2b. There is a strong positive relation, $r(6) = .98, p < .01$. The longer a person waited, the more frustration he or she experienced.

3. The last person to provide data appears to be an outlier, spending nearly twice as long in line as anyone else. If the relationship were graphed, this data point, would lie above the line that indicates the relationship for the other data points. This outlier, then, appears to be exaggerating the increase in frustration felt by people as they stood in line longer.

4a. $r = 0.86$.

4b. There is a positive relationship, $r(5) = .86, p < .01$. This indicates that as a therapist practices for more hours, his or her reported skill increases.

5a. $r = .58$.

5b. The relation between GRE scores and GPA in grad school is moderate but not significant, $r(8) = .56, p > .05$.

Module 36

Learning Objectives

- Understand the effect that sample size has on the strength of a correlation coefficient

- Distinguish between statistical significance and practical importance

- Understand that guidelines for practical importance differ depending on the use to which the correlation will be put

- Understand the effect that restriction in range has on the strength of a correlation coefficient

- Understand the effect that heterogeneity and homogeneity of the sample have on the strength of a correlation coefficient

- Understand the effect that instrument reliability has on the utility of a correlation coefficient

- Distinguish between correlation and common variance

- Distinguish between correlation and causation

Module Summary

- This module focuses on cautions concerning dealing with correlation. Correlations are among the most commonly used statistics, and it is important to be aware of how they can be misinterpreted.

- The first area of concern with correlational work involves sample size. The chances of rejecting the null are reduced as sample size shrinks; you may find variables that actually are related but have a relationship that does not appear significant because of a very small sample. Alternatively, you may find variables that do not appear to be related (or conceptually are not related) that show a significant correlation because of a very large sample. Ultimately, it is up to you to determine what findings are relevant.

- Similar to the previous point, a finding of statistical significance does not mean that the result is useful in practical terms. Unfortunately, there are no stringent guidelines to determine when a correlation is practical or impractical. Remember that a large enough sample size will cause any correlation coefficient to be significant. This means that it is up to you, the user of the statistical findings, to determine how strong a statistical relationship is necessary between two variables for you to consider the variables to be related strongly enough for that relationship to be useful to you. This will vary greatly depending on the situation. For example, as a teacher, you may desire a medium or smaller correlation between student grades on tests over time, as this would show that students are improving over time, as opposed to consistently earning the same grade (although that might be desirable, for students who are doing well). Alternatively, if you want to determine if two movie critics are providing similar opinions on movies (that is, if you are determining the reliability of their ratings), you may want a very high correlation.

- Another aspect to consider is restriction of range. *Restriction of range* is when the scores in your sample do not represent the full scale of variability for each variable. Imagine that you have given a test on which scores theoretically can range from 0 to 100. You want to see if scores on this test are related to final grades in the class. You select a random sample from your class and find a very small correlation ($r = .03$) between test scores and final grade in the class. You then look most closely at the test grades in your sample and discover that they range only from 70 to 73. This is a restriction of range. Your small correlation can be attributed to the lack of variability in test scores rather than a lack of a real relation.

- The next aspect to consider in a correlation is the homogeneity (similarity) and heterogeneity (difference) of those in your sample—your sample's diversity. This may influence your correlation because within your overall study, you have distinct subgroups. For example, you might assess the relationship between age and basketball performance among high school, college, and professional basketball players. It may appear that there is a strong positive relationship, with performance improving with age. This relationship actually can be attributed to the fact that basketball players don't become professional until they're relatively old, and the professional players generally exhibit higher levels of performance than the other groups. Within the group of professional basketball players, you may find a very different relationship between the variables of age and performance. Alternatively, homogeneity can decrease the correlation. If you were to look at the relationship between income and education of college professors, there may appear to be no relationship because they all earn similar amounts and have been in school for a similar amount of time.

- Another area of concern for correlational studies is the reliability of your measurements, or how well you are able to assess your variables of interest. Variables that are poorly measured should not be used as indicators. In doing correlational research, it is critical that you use accurate measurements so that you can draw appropriate conclusions about the relationships between variables. A common saying regarding this principle is "garbage in, garbage out": Poor measurements will result in poor conclusions.

- A common misconception is that the correlation coefficient represents the amount of shared variance between two variables. A correlation of .50 does *not*, however, mean that 50% of the variance in Variable A is attributed to 50% of the variance in Score B. This is because correlations are in linear units, whereas variances are in squared units. To determine the amount of shared variance, we must square the correlation coefficient. Thus, a correlation of .5 indicates that the variables share 25% (.5 ×.5) of their variance.

- The final and most common (and most egregious) mistake made using correlational data is assuming that a correlational relationship between the two variables implies a relationship of causation. Correlations do *not* imply causation. If two variables are related, that does not indicate that changes in one variable *cause* changes in the other. For example, there is a strong correlation between years of education and income; however, going to school for more years does not *cause* a pay increase. Instead, going to school longer will make a person more competitive in the job market and able to secure a higher-paying job. It is imperative that you avoid making this common error when interpreting correlational work.

True/False Questions

1. George notices that he becomes hungry around the same time each day. He concludes that the time of day is causing him to be hungry. George is correct in his thinking.

2. All statistically significant correlations indicate a meaningful relationship among variables and have "real-world" implications.

3. The proportion of shared variance between two variables is r^2.

4. Restriction of range decreases the strength of the correlation.

5. Sampling heterogeneity is likely to increase the strength of a correlation.

True/False Answers

1. False
2. False
3. True
4. True
5. True

Short-Answer Questions

1. How does an outlier affect a correlation?

2. How does sample size affect correlations?

3. A teacher asks students, at the end of every test, to rate the difficulty of the test. The teacher finds that students who rate the test harder tend to do better. Does this mean that a student who rates the test as very difficult will do better? Explain why or why not.

4. What is restriction of range? How does it affect correlations?

5. How do heterogeneity and homogeneity affect correlations? Why is this?

Answers

1. An outlier can reduce the strength of a correlation if the outlier falls outside the linear trend created by the other scores.

2. Large sample sizes reduce the critical value needed to declare a result statistically significant. This is because the probability of observing a consistent relation between two unrelated variables becomes lower; thus, larger samples have lower critical values. Alternatively, the chances of observing some type or relation between only a handful of scores is substantially higher. Thus, smaller samples have higher critical values.

3. No, this does not indicate that giving the test a higher rating difficulty rating will cause a higher grade. This is an error of inferring causation from a correlation. It may be the case that students who rate the test as difficult were aware of the complexities of the questions and had to think a great deal to discover the correct answer. Students who did not rate the test as difficult may not have prepared heavily for the test and thus may have misperceived the test as simple.

4. A restriction of range is when a data set is missing scores from a portion of the scale of measurement. This may limit the interpretability of the relationship.

5. Heterogeneity of the sample can increase the strength of the correlation. Heterogeneity implies that there are distinct subgroups within the sample that have different levels of

"ability" on the two variables. This can create the false appearance of a strong relation between the two variables. In contrast, homogeneity of the sample can decrease the correlation. This is attributed to the participants in the sample being so similar that the sample fails to use the full scale of measurement.

Multiple-Choice Questions

1. A researcher obtains a correlation of .6 but finds that the result was not significant. The relationship seems to have great practical importance, and he believes the study has led to a Type 2 error. What could the researcher do to obtain statistical significance?

    a. Increase the sample size

    b. Use different measures

    c. Change the hypotheses of the study

    d. There is nothing the researcher can do

2. One study found that the number of people who drowned per month was significantly correlated to ice cream sales. A friend of yours discovers this finding and tells you not to buy ice cream because it will lead to drowning. Which pitfall of correlation has your friend succumbed to?

    a. Homogeneity of sample

    b. Restriction of range

    c. Correlation implying causation

    d. Common variance

3. Which of the following describes an outlier?

    a. Increases the strength of a relationship

    b. Decreases the strength of a relationship

c. Can be attributed to a data entry error

d. B & C

4. What pitfall of correlation does the "garbage in, garbage out" statement refer to?

a. Outliers

c. Common variance

b. Unreliable measures

d. Correlation and causation

5. Peggy is a schoolteacher and wants to determine if the ages of the students in her kindergarten class are related to their reading ability. Here are the data she obtained. Which pitfall should Peggy be most mindful of in doing this study?

Age	Score
5.3	94
5.6	95
5.4	92
6.1	89
5.9	99

a. Homogeneity of her sample

c. Correlation implying causation

b. Outliers

d. Imprecise measures

6. A researcher finds that the correlation between stress and intensity of migraine pain is .75. What is the proportion of shared variance among migraine pain and stress?

a. .75

c. .25

b. .56

d. .87

7. Two variables share 32% of their variance. What is the Pearson r for these variables?

a. .32

c. .10

b. .57

d. .64

Multiple-Choice Answers

1. A

3. D

2. C

4. B

5. A   7. B

6. B

Module Quiz

1. The relationship between GRE scores and grad school GPA was found in one study to be not
    significant statistically. Looking at the data below, what pitfalls do you notice that may have
    influenced this relationship?

GRE	GPA in Grad School
1160	3.25
1180	3.14
1200	3.78
1280	3.9
1290	3.8
1370	4.0
1400	4.0
1520	4.0
1570	3.9
1590	3.7

2. Why does correlation not imply causation?

3. The head of a university is interested in determining how well teacher competence, as
    measured by a standardized test (rated 0-100), is related to student satisfaction in the
    classroom (rated 0-10). The following data are obtained from 5 teachers. The correlation
    associated with the following data was found to be not statistically significant. What pitfalls
    might have reduced the strength of this correlation?

Competence	Satisfaction
84	10
57	10
85	9
41	10
43	2

4. A correlation between two variables is found to be .57. Does this mean that the two variables share 57% of the variance? Why or why not?

5. How may increasing the heterogeneity of a sample improve the strength of the relationship?

Quiz Answers

1. There are two potential pitfalls with this sample. First, there appears to be a restriction of range for GPA. Second, the sample is small.

2. Significant relationships between two variables do not necessarily indicate that one variable causes the other. The precise causal relationship between the two variables is best determined by using additional knowledge about what variables are being correlated. Two variables might be related because they are both caused by a third variable, for example, and the fact that two variables are correlated does not necessarily say anything about which one is the "cause" and which one is the "effect."

3. The sample size is very small, so the correlation coefficient would not be reliable. A restriction of range for the student satisfaction variable reduced the strength of the correlation; almost all the students polled reported scores of 9 or 10, regardless of their rating of teacher competency. Finally, the measuring instruments for teacher competency and student satisfaction might not be reliable.

4. The two variables do not share 57% of the variance. Common variance between two variables is measured by the square of the correlation. The shared variance for these variables is 32%.

5. Increased sample heterogeneity can increase the strength of the relationship between two variables by increasing the range of scores.

Module 37

Learning Objectives

- Understand why a line of best fit minimizes error of prediction

- Find the equation for the prediction line for a set of bivariate data

- Predict scores on Variable Y from known scores on Variable X

Module Summary

- Recall that one of the uses of correlation is prediction—using a person's score on Variable A to estimate what his or her score would be on Variable B. In this situation, we would refer to Variable A as the ***predictor*** and Variable B as the ***criterion***. When the predictor perfectly predicts the criterion, the scatterplot will resemble a straight line and the $r = 1$ (or -1). In this situation, we can determine any person's score on Variable B as long as we know that person's score on Variable A. When a score on Variable B (the Y variable) is predicted by a score on Variable A (the X variable), the predicted score is referred to as **Y** .

- In the social sciences, variables are rarely perfectly correlated. Thus, when using a score on X as a predictor, there may be many different possibilities for a score on Y. This poses a problem, as we can have only a single criterion score per predictor score. Probability is used to solve this situation, in that we state that the criterion will predict the most probable single score. All of these most probable criterion scores for each predictor score form a straight line through the center of all the points of a scatterplot. This line is referred to as a ***regression line*** or a ***best fit line***.

- As mentioned, not all of the scores in the scatterplot will fall on the regression line. Sometimes a Y value will be above or below Y . The difference between a Y value and its Y $(Y - Y$ $)$ is referred to as ***prediction error***. This is a measure of how far off the prediction was from the

370

actual value. Similar to the function of a mean, the summed value of the deviations above the prediction score is equal to the summed value of deviations below the prediction score.

- Prediction error plays a role in selecting where the best fit line will be drawn. The best fit line minimizes the amount of prediction error across all points. The conceptual method that we use to determine how the best fit line will be drawn is as follows: (1) Find the deviation of each raw Y from Y . (2) Square the prediction errors. (3) Sum the squared prediction errors. (4) Do this for all possible lines. (5) Select the line for which the sum of squared prediction errors is the smallest. This method is referred to as the **least squares** method because we are selecting the line that has the smallest (least) sum of squared prediction error.

- The mathematical method of calculating a regression line uses the same formula as that for a straight line: $y = mx + b$, where $m=$ the slope and $b =$ the intercept on the Y-axis. Slope refers to the unit change in Y for each unit change in X. A slope of 1 indicates that an increase of one unit on X is related to an increase of one unit on Y. The formula that is commonly used in statistics is $Y = bX + a$, where $b =$ the slope and $a =$ the intercept on the Y-axis. Using our knowledge of the relationship between two variables in this equation, the formula changes to:

$$Y' = r_{XY}\left(\frac{s_y}{s_x}\right)X - r_{XY}\left(\frac{s_y}{s_x}\right)M_X + M_Y$$

- The previous equation, although more intimidating, is equivalent to $Y = bX + a$. Once actual values are entered and the equation is reduced, the first term represents the slope and the second represents the Y-intercept.

Computational Exercises

1. Using a regression equation for which $b = 0.13$ and $a = 0.33$, what is the value for Y for the following values of X? a) 3, b) 7, c) 5.43?

2. Using the regression equation in question 1, what is the prediction error for each of the scores if the observed Y values were, respectively, a) 4, b) 6.32, c) 3.21? Which Y has the smallest amount of prediction error?

3. After attending the birthday parties for a 5-year-old and then another for a 25-year-old, you are interested in determining the relationship between age (X) and the numbers of birthday presents (Y) a person receives. You discover that the correlation is .32, using the following data:

	Age	Presents
Mean =	15.6	5.4
s =	7.3	3.9

What is the regression equation associated with this information? How many presents should you expect on your next birthday?

4. Below are data obtained from a study that examined the prevalence of PTSD in a population of war refugees. Researchers were interested in determining the extent that exposure to war trauma (X) was related to the severity of the PTSD symptoms (Y). They found a correlation of .23 with the following data:

	Exposure	PTSD Severity
Mean =	4.7	11.23
s =	2.3	1.90

What is the regression equation associated with this information?

5. A developmental psychologist is interested in determining if the number of hugs (X) a child receives per day is related to the amount of affection (Y) the child feels from his or her parents. The psychologist obtains a correlation of .65 and the following data:

	Hugs	Affection
Mean =	7.8	32.8
s =	4.6	7.69

How much affection should a child who receives 12 hugs a day feel toward his or her parents?

6. A parent notices that her adolescent appears to be in a new relationship each week and that her adult child has been in the same relationship for the past few months. She decides to conduct a

372

study to determine if age is related to the length of a relationship in months. Using data from many teens and adults, she finds a relationship of .35 with the following data:

	Age	Length of Relationship
Mean =	25.0	3.70
s =	5.4	0.89

How long should a relationship last for a person who is 15? A person who is 28?

7. The parent in question 6 finds a 15-year-old who has been involved with someone for the past 8 months and a 28-year-old who has been involved with someone for only 0.3 months. What are the prediction errors for these scores? Based on these observations, how reliable is this line?

Computational Answers

1. a) 1.12; b) 2.44; c) 1.96.

2. a) $4 - 1.12 = 2.88$; b) $6.32 - 2.44 = 3.88$; c) $3.21 - 1.92 = 1.29$. The last score (c) has the smallest amount of prediction error.

3. $Y = 0.17X + 2.73$. The number of presents you should expected depends upon your age. Here is the number of presents for someone of age 21: $Y = 0.17(21) + 2.73 = 6.30$.

4. $Y = 0.19X + 10.34$.

5. $Y = 1.09X + 24.32$. The child should have an affection rating of 37.40.

6. $Y = 0.06X + 2.25$. A 15-year-old should have a relationship lasting 3.15 months; a 28-year-old should have a relationship lasting 3.93 months.

7. 15-year-old $= 8 - 3.15 = 4.85$ months.

28-year-old $= 0.3 - 3.93 = -3.63$ months.

The prediction error is very large, so the regression equation (line) appears to be not very reliable.

True/False Questions

1. Linear regression lines can be calculated for curvilinear relationships.

2. A predictor variable causes the criterion variable.

3. Comparing algebraic notion to statistical notation, $m = a$ *and* $a = b$.

4. The slope of a regression line is affected by the strength of the relationship between the two variables.

5. The means of each variable are used to calculate the constant of the regression equation.

6. Regression lines work best for normally distributed data.

7. For the regression equation $Y' = 21.3X + 312$, $Y' = 312$ when $X = 0$.

8. The regression line is a series of point estimates for each value of Y based on each value of X.

True/False Answers

1. False	4. True	7. True
2. False	5. True	8. True
3. True	6. True	

Short-Answer Questions

1. What is a regression line?

2. Why is the regression line like a series of means?

3. What is the least squares method?

4. How does shared variance affect the position of the regression line?

5. How should regression lines be used for skewed data?

6. How does sample size affect a regression equation?

7. A predictor variable is used to estimate a score on a criterion variable. Does this mean that a predictor variable causes the score on a criterion variable? Why or why not?

<u>Answers</u>

1. A regression line, or best fit line, is a formula that is used to predict a score on Y using a score on Y.

2. A regression line is considered a series of means because the sum of the deviation scores above the regression line is equal to the sum of the deviation of score below the regression line. Also, each point along the line is a mean Y score for each X score.

3. The least squares method is the theoretical backbone of a regression line. It states that the regression line is the one that provides the overall smallest (least) sum of squares for the prediction errors.

4. As the amount of shared variance (and, hence, correlation) increases, the slope of the regression line increases.

5. When dealing with skewed data, it may be more appropriate to use multiple regression lines to better predict the criterion variable.

6. Sample size does not directly affect a regression equation; however, larger samples can provide more accurate estimates of correlations. With a more accurate correlation coefficient, the regression line will provide a more accurate prediction of Y from X.

7. No, it does not. This statement is the error of correlation implying causation.

Multiple-Choice Questions

1. How many Y values are there per X?

      a. 1   c. 3

      b. 2   d. 4

2. How many Y values are there per X?

      a. 1   c. 3

      b. 2   d. Depends on the data

3. For which type of data is a single regression line inappropriate?

    a. Positively skewed

    b. Negatively skewed

    c. Normal

    d. A and B

4. What is the slope of a best fit line with an $r = 0.75$, $s_x = 2.1$, and $s_y = 4.9$?

    a. 0.32

    b. 1.75

    c. 2.33

    d. 0.43

5. SAT scores and college GPA have a correlation of .25. GRE scores and graduate school GPA have a correlation of .13. Without any additional information, which regression line should you expect to be more accurate?

    a. SAT and college GPA

    b. GRE and graduate school GPA

    c. Both are equal

    d. It depends on the Y-intercept

6. The correlation between the amount of time spent talking on the phone and the size of the phone bill is .73. Using the additional information below, what is the estimated bill of a person who talks for 320 minutes?

	Minutes	Bill
Mean =	450	98.65
s =	57	6.35

    a. 62.05

    b. 25.6

    c. 87.79

    d. 149.70

7. Using the information from the previous question, what is the estimated bill of a person who talks 213 minutes?

    a. 65

    b. 87.65

    c. 79.23

    d. 62.05

8. A salesman is trying to impress his customers by accurately guessing their waist sizes based on their height. The correlation between height and waist size is .78. He bases his assumptions on the following data taken from the store's sales history.

	Height	Waist Size
Mean =	68	30.0
s =	8	2.8

What is the regression equation that the salesman uses?

a. $Y = .27X + 11.44$

c. $Y = .78X + 15.98$

b. $Y = 1.27X + 12.44$

d. $Y = 1.3X + 98.7$

9. The salesman from the previous example deeply offends a customer by overestimating his waist size. The customer was 78 inches tall, with a waist size of 28. How far off was the salesman?

a. 2.5 inches

c. 9.5 inches

b. 4.5 inches

d. 11.2 inches

Multiple-Choice Answers

1. A

6. C

2. D

7. C

3. D

8. A

4. B

9. B

5. A

Module Quiz

1. Below are the data taken from a nutrition study that attempted to assess the relation between daily calorie consumption and weight. The correlation was found to be .54. Using the additional data provided below, what is the estimated calorie consumption for someone who weighs 189 pounds?

377

	Weight	Calorie Consumption
Mean =	200.00	1800
s =	30.87	300

2. You notice that the cost of clothing in private boutique stores is much higher than in general department stores. You assume that this is because clothing designers spend more time creating each piece of clothing than those in department stores. If the relationship between cost of clothing and the time spent creating it (in days) is .63, what is the regression equation using the following information?

	Cost	Time of Creation
Mean =	78	34.0
s =	9.45	4.5

3. A farmer is interested in determining how many pounds of fertilizer he should use on his potato crop to have it grow to about 0.50 pounds per potato. The relationship between potato size and fertilizer usage is .21. Using the following data taken from numerous farms around the country, how much fertilizer should he use?

	Potato Size	Fertilizer
Mean =	0.47	8.7
s =	0.12	0.16

4. You notice that many of your friends who love music have a large collection of digital music. If the correlation between a person's self-rated love of music and the amount of digital music they own is .89, what is the regression equation based on the following data?

	Love of Music	Digital Music
Mean =	6.5	45.9
s =	2.9	7.84

5. A neuroscientist is interested in determining how anxiety is related to the amount of GABA in a person's system. He conducts a study and determines that the relationship between

GABA and anxiety is –.72. What is the regression equation used to predict how anxious a person will be based on the amount of GABA in his or her system?

	GABA	Anxiety
Mean =	87.9	25.47
s =	20.4	10.23

Quiz Answers

1. $Y = 5.25X + 750.44$. A person who weighs 189 pounds is predicted to consume 1742.69 calories per day.

2. $Y = 0.3X + 10.60$.

3. $Y = 0.28X + 7.38$. The farmer should use 8.78 pounds of fertilizer.

4. $Y = 2.41X + 30.24$.

5. $Y = -0.36X + 57.21$.

Module 38

Learning Objectives

- Know the shape of a distribution of prediction errors

- Calculate a standard error of prediction

- Establish intervals having various probabilities of containing the predicted score

- Understand the effect of sample size on the size of the standard error of prediction

- Understand the effect of variability among predicted scores on the size of the standard error of prediction

Module Summary

- This module focuses on calculating confidence intervals for regression lines. A regression line can be considered a number of Y score estimates, one for each X score; however, there may be a large range of actual values for each Y at each X, and thus it can be helpful to determine where these actual scores are expected to fall. Recall that the stronger the correlation, the less the data deviate from the prediction line (the smaller the prediction error). In contrast, the weaker the correlation, the more the data will deviate from the prediction line.

- The prediction error for any particular Y variable will be normally distributed. Most scores will deviate a small amount from Y , and very few will deviate a great deal. The extent to which a Y score deviates from Y can be transformed into standard deviation units, which are referred to as the ***standard error of prediction***. This is the average amount of linear dispersion among the units that are plotted. The standard error of prediction is calculated with the following formula:

$$s_{YX} = s_y \sqrt{1 - r_{XY}^2}$$

- Now that we have a standard deviation value for the expected difference from Y, we can calculate a confidence interval. The confidence interval can be calculated using the same proportions as those for the normal curve. In other words, about 68% of the scores will fall within 1 standard error, about 95% will fall within 2 standard errors, and so on.

- Only two variables are able to influence the size of the standard error. These are the size of the standard deviation for the Y variable and the strength of the correlation between X and Y. As the strength of the correlation increases, the standard error decreases. Similarly, as the standard deviation of Y increases, so does the standard error.

Computational Exercises

1. The standard deviation of Y is 3.25. What will the standard error of prediction be if the correlation between X and Y is 0.407? What will it be if the correlation between X and Y is 0.80?

2. You notice that the cost of clothing in private boutique stores is much higher than in general department stores. You assume that this is because clothing designers spend more time creating each piece of clothing than those in department stores. The relationship between cost of clothing and the time spent creating it (in days) is .63, with a best fit line of Y = $0.3X + 10.60$. Using the following information, what is the 95% confidence interval for the time of creation for a dress that costs $132?

	Cost	Time of Creation
Mean =	78	34
s =	9.45	4.5

3. After attending the birthday parties for a 5-year-old and then another for a 25-year-old, you are interested in determining the relationship between age (X) and the number of birthday presents (Y) a person receives. You discover that the relationship is .32, with a best fit line of $Y = 0.17X + 2.73$.

	Age	Presents
Mean =	15.6	5.4
s =	7.3	3.9

What is the 95% confidence interval for the 5-year-old? What is the 95% confidence interval for the 25-year-old? After calculating your answers, comment on anything that surprises you.

4. A neuroscientist is interested in determining how anxiety is related to the amount of GABA in a person's system. She conducts a study and determines that the relationship between GABA and anxiety is –.72, with a best fit line of $Y = -0.36X + 57.11$. If the $s_y = 10.23$, what is the 99% confidence interval of values to be added to and subtracted from each predicted Y?

5. A farmer is interested in determining how many pounds of fertilizer he should use on his pumpkin crop. The relationship between pumpkin size and fertilizer usage is .21, with a best fit line of $Y = 2.80X + 7.38$. If the $s_y = 1.60$, what is the 95% confidence interval to be added to and subtracted from each predicted Y?

Computational Answers

1. When the correlation is 0.40, the standard error of prediction is 2.98. When the correlation is 0.80, the standard error of prediction is 1.95.

2. Standard error of prediction = 3.49. For a dress that costs $132, the 95% confidence interval would be 132 +/– (6.98) = 125.02 to 138.98.

3. For the 5-year-old: 3.58 +/– (2) (3.695) = 3.58 +/– 7.39 = –3.82 to 10.97.

For the 25-year-old: 6.98 +/– (2) (3.695) = 6.98 +/– 7.39 = –0.41 to 14.37.

Surprises: (1) A linear exists relationship between age and gifts, with more gifts given to older people than to younger people. (2) The standard error of measurement is so large that the intervals include negative amounts gifts!

4. The 99% confidence interval value to be added to and subtracted from each predicted Y is 21.30.

5. The 95% confidence interval value to be added to and subtracted from each predicted Y is 3.13.

True/False Questions

1. Stronger correlations reduce the overall amount of prediction error.

2. A best fit line is the one with the smallest sum of squared deviations for each Y value.

3. Least squares refers to using the regression equation with the least amount of error per X.

4. Better measurements of variables correspond to more accurate prediction based on a regression line.

5. Y values are thought to be normally distributed about their corresponding Y .

6. The standard deviation of X plays a large role in the standard error of the estimate.

True/False Answers

| 1. True | 3. True | 5. True |
| 2. True | 4. True | 6. False |

Short-Answer Questions

1. What is the standard error of the estimate?

2. How are confidence intervals calculated for a Y value? Why is this method used/allowed?

3. How does the strength of a correlation affect the standard error? Why is this?

4. How will using a Y score that has a tremendous amount of variability affect the ability of an X score to act as a predictor?

5. One of the pitfalls of using a correlation concerned using poor measures, as they provide poor predictors of other variables. Using your new knowledge of regression lines and prediction error, how would poor measures affect the placement of a regression line?

6. Recall that outliers (extreme scores) can weaken the relationship between two variables. Using a regression line, how would an outlier affect a regression line?

7. Explain why the Y values around each individual Y' score are expected to be normally distributed.

Answers
1. The standard error of the estimate is the average deviation of a score on Y from its Y'.

2. Confidence intervals are calculated by multiplying the square root of (1 minus the correlation between X and Y squared) by the standard deviation of Y. This is allowed because it is expected that Y scores will be normally distributed around each Y'.

3. An increase in the strength of the correlation decreases the size of the standard error. This is because a stronger correlation indicates that the scores will be bunched more tightly around the regression line.

4. It will increase the standard error of prediction and thus reduce the ability of X to predict scores on Y.

5. Measures with lots of prediction error will "flatten out" the regression line, making the line appear more horizontal. This is because measures with prediction error can be expected to have weaker correlations and thus a more gradual slope. In addition, measures with lots of prediction error will cause the data to be more widely distributed (more distant) around the regression line (or the line will fall within a wider swarm of points).

6. Outliers would pull the regression line in the direction of the outlier. This will directly affect the slope of the regression line.

7. $\hat{Y}$ is thought to the be most probable score for Y based on X; therefore, it is expected that small deviations from $\hat{Y}$ are more common than large deviations. This pattern forms a normal distribution.

Multiple-Choice Questions

1. As the strength of the correlation increases, prediction error _____.

   a. Increases   c. Remains the same

   b. Decreases   d. Depends on the data set

2. As the standard deviation of Y _____, the accuracy for predicting Y from $\hat{Y}$ _____.

   a. Increases, increases   c. Increases, decreases

   b. Decreases, decreases   d. Increases, remains the same

3. The correlation between the amount of time spent talking on the phone and the amount of the phone bill is .73, and the $s_Y = 6.35$. What is the standard error of prediction?

   a. 65.78   c. 1.36

   b. 8.65   d. 4.34

4. Using the information and answer from the previous question, what would the 95% confidence interval be for a person who spoke 156 minutes?

   a. 150.8 to 187.6   c. 147.3 to 164.7

   b. 89.65 to 90.32   d. 134.7 to 152.9

5. One of the factors that is related to the size of the standard error is

   a. The relationship between X and Y c. The mean of Y

   b. The mean of X d. The standard deviation of X

6. A salesman is trying to impress his customers by accurately guessing their waist sizes based on their height. The correlation between height and waist size is .78, with $s_y = 2.8$. This leads to him offending a customer with a 28-inch waist because of his inaccurate prediction. To console him, what was the standard error of prediction for this regression equation?

    a. 9   c. 1.75

    b. 10.23   d. 3.75

7. After finding out the standard error, the salesman doesn't feel much better. In your last attempt to console him, what would the 99% confidence interval be for a person with a 28-inch waist?

    a. 15.32 to 20.85   c. 30.75 to 34.25

    b. 19.6 to 36.4   d. 22.98 to 27.45

Multiple-Choice Answers

1. B   5. A

2. C   6. C

3. D   7. B

4. C

Module Quiz

1. You notice that many of your friends who love music have large collections of digital music. If the correlation between a person's self-rated love of music and the amount of digital music they own is .89 and the $s_y = 7.84$, what is the 99% confidence interval for each Y?

2. A nutrition study demonstrated that there is a correlation of .64 between daily calorie consumption and weight. If the $s_y = 6.78$, what is the 95% confidence interval for each Y?

3. If variables X and Y are completely uncorrelated ($r = 0$) and Y has a standard deviation of 10, what is the standard error of prediction expected to be?

4. Using the regression equation of $Y = 2.4X + 12.1$, what is the standard error of the prediction associated with a $Y = 5$ and $X = 4$ where $r = 0.56$ and $s_Y = 12.5$?

5. What are the two factors that influence the standard error of prediction?

Quiz Answers

1. The 99% confidence interval is 10.72 around the predicted value.

2. The 95% confidence interval is 10.42 around the predicted value.

3. The standard error also would be 10.

4. $s_{YX} = (12.5)(\sqrt{1-(0.56^2)}) = (12.5)\sqrt{0.687} = 10.36$.

5. The standard error of prediction is influenced by the standard deviation of Y and the correlation between X and Y.

Module 39

Learning Objectives

- Decompose total variance into its components

- State the advantage of multiple predictors over a single predictor

- Use predictor scores and their coefficients to predict outcomes

- Explain the logical difference between experimental and regression analytic methods

- Locate and interpret key statistics found on multiple regression software output

Module Summary

- A primary goal of statistics is to improve the accuracy of prediction. Up until this point, we have learned how to predict one variable (Y) using another variable (X) with best fit lines and correlations. The estimates that we can obtain when using a single variable (X) to predict another variable (Y) can be inaccurate. The lower the correlation between X and Y, the weaker the prediction.

- To increase our powers of prediction, we can include additional variables (X's). In other words, rather than using GPA alone to predict SAT scores, we could use GPA and number of hours spent studying for the SAT. By including the additional variable of hours spent studying, we would be able to better predict a person's SAT score. The use of multiple predictors (X's) to estimate a single Y is called ***multiple regression***.

- A multiple regression follows the same logic as simple regression. The formula for a single regression was $Y = bX + a$, where b represents the slope for a single X. For multiple regression, we just add additional X variables and their slopes so that the equation looks like this:

$$Y = b_1X_1 + b_2X_2 + b_3X_3 + \ldots b_nX_n + a$$

388

- The use of multiple b's slightly changes the meaning of each of them. They are now considered **weights**, meaning that they determine how strong (or heavy) a predictor is in its relation to Y . If a variable has a very small b, then it is believed to have a weak individual relationship with Y , holding the other variables constant. Alternatively, a large b is believed to indicate a strong relationship between its X variable and Y .

- If you want to compare the difference between two regression weights in a single equation, you should use the **beta** weights, or B. These represent the standardized regression weights between two variables. As always with standardization, comparisons can be made across different scales with variables when using the standardized values, but not when using the raw score values.

- When using multiple regression, we determine the **common variance** across all of the predictors of Y . In simple linear regression, this was obtained by calculating r^2 for the single predictor. Squaring the B for each predictor, however, will result in a common variance greater than 1.00. This is because all of the predictors are interrelated—the X's are correlated with each other as well as with Y. We need to account for the interrelations among the X's to determine their common variance with Y.

- Accounting for common variance of multiple X's and a single Y can be an arduous process that involves calculating correlations between each X and Y and then between each pair of X's. As a result, it is best to use a statistics software program to estimate these relations. By including all of your X's in your multiple regression, you would obtain an R^2, the total common variance that all of the X's share with Y. This value will be on a scale of 0 to 1.

- Multiple regression enables us to answer two important questions: (1) Do all of our predictors taken together account for a significant amount of variability in Y? and (2) Does

389

each individual X predictor account for a significant amount of variability in Y? You can use Cohen's values to determine if the size of the effect for all variables is medium ($r = .30$, $R^2 = .09$) or large ($r = .40$, $R^2 = .16$). The second question is answered by the standardized regression weights (B's). The individual B's can also be evaluated with a t test to determine if each contributes to the criterion individually.

- When we include multiple predictors in a multiple regression, we have developed a *model*—an estimate of how we think we can best predict a criterion Y from several predictors (X's). Another feature of multiple regression is the use of *model testing*. This occurs when we compare how accurate some of our predictions of Y are when we use only a few of our X's (or a *reduced model*) with how accurate the predictions are using all of our X's (a *full model*).

- Multiple regression is yet another form of the prototype formula that has been emphasized throughout the book, of "What did you get?" versus "What did you expect to get?" In multiple regression, the formula becomes our predicted variance (obtained from our X's) over the total variance, rather than over the error variance. The similarity across all of the techniques we have learned thus far and multiple regression is not by chance! All of these approaches are part of a set of approaches referred to as the *General Linear Model (GLM)*. The primary feature of the GLM is that because treatment (independent variable) variance is divided by total variance, the statistic can be interpreted directly as a percentage or a proportion of the total available variance accounted for by the treatment variables.

True/False Questions

1. R^2 can exceed a value of 1 because there are multiple variables in the regression equation.

2. R^2 is a measure of effect size in multiple regression that examines the amount of variance accounted for by all of the predictors.

3. A full model will never predict more variance than a reduced model.

4. The residual variance for a model with $R^2 = 0.13$ is 0.87.

5. The GLM is a rarely used technique that focuses on only a small subset of analyses.

True/False Answers

1. False   3. False   5. False

2. True   4. True

Short-Answer Questions

1. How does multiple regression differ from simple regression?

2. What are the advantages of including additional predictors in a regression equation?

3. Explain the difference between a b and a beta (B) in multiple regression.

4. What are the two main hypotheses that are tested in a multiple regression?

5. What information is provided in the ANOVA in a multiple regression?

Answers

1. Multiple regression uses multiple variables to predict Y , whereas simple regression uses only a single variable.

2. By including multiple predictors, multiple regression provides a more accurate estimate of Y'.

3. B's refer to the standardized regression weight, whereas the b refers to the unstandardized regression weight. The standardized regression weight can be compared across variables to determine which variables are the strongest predictors of Y'. The unstandardized regression weight provides a measurement of how much Y' changes per unit increase in an X variable.

4. The first is "Do all of the predictors account for a significant amount of variance in the criterion?" and the second is "Does each individual predictor account for a significant amount of variance in the criterion?"

5. The ANOVA provides another method to determine if the predictors account for a significant amount of variance in the criterion. It tells us if the R^2 is significant.

Multiple-Choice Questions

1. What is the residual variance for a study with an R^2 of .89?

    a. 0.42

    b. 0.11

    c. 0.32

    d. 0.89

2. Which model has a greater residual variance—one with an R^2 of .23 or one with a residual variance of .64?

    a. One with $R^2 = .23$

    b. One with a residual variance of .64

    c. Both have equal residual variances

    d. You need more information

3. In a model with 4 predictors and R^2 of .16, each predictor accounts for how much variance?

    a. .04

    b. .40

    c. .16

    d. You need more information

4. A study has 5 predictors in the model. Using the rule of thumb for sample size, at least how many participants should be in this study?

    a. 50

    b. 25

    c. 30

    d. 40

5. The standardized coefficient for Predictor A in a model is .32 and for Predictor B is .12. Which of these variables is more strongly related to the criterion variable?

    a. Predictor A

    b. Predictor B

    c. Both are equally related

    d. You need more information

Multiple-Choice Answers

1. B

2. A

3. D   5. A

4. A

Module Quiz

1. How would you expect the R^2 of a model to change by removing a predictor variable? By

    adding in a predictor variable?

2. How does the order in which predictors are added to the model impact the total R^2 value?

3. What value should you use to determine how much of a change is expected in a criterion

    variable based on a unit increase in a predictor variable?

4. The unstandardized coefficient for Predictor A is 1.32 and for Predictor B is 0.31. If the

    constant is 93.12, what is Y' if A = 38 and B = 12?

5. Use the following output to determine what proportion of total variance was accounted for by

    the predictors.

Model Summary

Model	R	R Square	Adjusted R Square	Std. Error of the Estimate
1	0.588[a]	0.346	0.280	7.42851

a. Predictors: (Constant), Avoidance, prLSASfear

Quiz Answers

1. When adding a predictor variable, the R^2 is expected to increase. When removing a predictor

    variable, the R^2 is expected to decrease.

2. The order in which predictors are added will not affect the final R^2 value. The R^2 that is

    affected by order is that of the steps.

3. The b provides this value.

4. $Y' = (38*1.32) + (12*0.31) + 93.12 = 147.$

5. The predictors accounted for 35% of the total variance ($R^2 = .346$).

Module 40

- You have now learned many different statistical procedures that enable you to answer a variety of questions related to data sets. Unfortunately, real research questions are not presented as neatly they have been in this book. The biggest challenge when preparing to do an analysis can be determining which type of analysis will best answer your research questions.

- In determining how to go about selecting the best test, it can be helpful to ask yourself questions about the characteristics of your research question. These include questions about the study's purpose (displaying or reporting information, establishing a relationship, establishing a cause), the scales of measurement that are used, the number of groups you have in the data, and the number of variables used. The first flowchart in this module provides a good model for the different questions you should ask regarding your analyses.

- Recall that descriptive statistics are used to describe a group of numbers. One of the simplest ways to describe a group of numbers is by creating a table or graph that displays the entire data set. A frequency table shows the number of cases that occur for each given score; it can be used to show proportions of the data that fall within various ranges of scores as well as percentages in relation to the total sample. Graphs provide a visual depiction of the data and enable researchers to determine the general shape of the distribution of data. For single variables, it can be helpful to create a histogram, which displays the range of scores on the X-axis and the frequency of the scores on the Y-axis. For two variables, a scatterplot is used. Individual scores are represented by individual dots that correspond to the values for a data point for the variables X and Y.

- Other descriptive statistics can summarize entire sets of data in a few numbers. Some of the most commonly used descriptive statistics are measures of central tendency: the mean, median, and mode. A mode is the most frequently occurring score, the median is the score that occurs at the exact center of the distribution, and the mean is the average score in the distribution. In addition to being the average number, the mean is also the "balance point" of the data. Remember that in a normal distribution, all of the measures of central tendency are equal.

- Dispersion summarizes the extent of variability within the scores in a data set. These measures include the range (the difference between the highest and lowest score), the variance (the average area distance from the mean), and the standard deviation (the average linear distance from the mean). The large majority of scores in a normal distribution occur within 3 standard deviation on each side of the mean.

- Raw scores can be converted into standard deviation units. These scores are referred to as standard scores. Standard scores (or z scores) tell us how many standard deviations a raw score is from the mean of the distribution. Also, the sign of the standard score (+/−) specifies where the score is in relation to the mean (above or below it).

- A correlation coefficient provides information about how two variables related. From a correlation coefficient, we can determine if the variables are positively related or negatively related, as well as the strength of this relationship. It is important to note that no matter how strongly two variables are related, a correlation does not indicate that the relationship between the two variables is causal.

- The other types of statistics that we focused on were inferential statistics. Inferential statistics are used to draw conclusions about a population from sample data. Inferential statistics usually are focused on determining if one treatment group differs from another group. The multiple different inferential tests that can be done are shown in the second flowchart in this module.

- Parametric tests compare sample statistics to population parameters. A one-sample test compares a sample mean to a population mean. If we know the population standard deviation, we can use a normal deviate z test. If not, we use a one-sample t test. For either test, if the difference between the sample mean and population mean is greater than a predetermined amount, established by the probability of making a Type 1 error and by the standardized random error, then we can assume that the difference is not attributable to chance, and the sample is representative of a different population. A two-sample test assesses the difference between the means of two samples, one that has been given some type of treatment and one that has not, in a manner similar to that of a one-sample test. Finally, a multisample test simultaneously addresses the differences between multiple groups. A one-sample ANOVA is used when there is only one independent variable with multiple groups. A second independent variable can be used to assess interaction effects. This is referred to as a factorial ANOVA. The purpose of an ANOVA is to compare between-group differences (how different the groups are from one another) to within-group differences (how different the individuals within each group are from one another). If the between-group differences are more than expected (greater than within-group differences), then the groups are considered to be significantly different. After obtaining a significant result, a post hoc test is conducted to determine which groups (treatments) are significantly different. Two types of post hoc test are the Tukey HSD and the Scheffé test.

- Correlation coefficients tell us not only the strength and direction of the relationship between two variables, but also the probability of observing that relationship between two variables. If the probability of observing the obtained relationship by chance alone is low, then we can conclude that the variables are in fact significantly related. The strength and direction of this relationship is provided by the correlation coefficient.

- Nonparametric tests address variables that are not based on population parameters. The questions we are able to ask with nonparametric tests include those about ranking across groups, similarities in the shapes of data distributions, and response patterns. The first nonparametric test that was covered was the chi-square test for goodness of fit. This test determined how well the frequencies in one group fit (or are similar to) that of another group. The other nonparametric test that was covered was the chi-square test of independence. This test is similar to the interaction term of a factorial ANOVA, in that it seeks to determine if two variables act independently of one another. If the two variables are independent, the expected frequencies among the different categories are expected to be proportional to the variables' relative frequencies. If the observed frequencies are substantially different from the expected values, then we conclude that the independent variables are not independent.

Computational Exercises

For each of the following, first identify the hypothesis (if any) that needs to be stated before answering the question.

1. The following data are taken from a sample of statistics students who were asked to rate their enjoyment of statistics, on a scale of 1-100. Describe these data to a colleague using a measure of central tendency, the standard deviation, and the shape of the distribution.

Score		
99	94	36
9	77	8
49	27	65
97	32	32
68	62	62

2. You obtain a sample of responses from students in your current statistics class regarding their enjoyment of the course, shown below. Describe the data in a similar manner to that of the previous set of responses (mean, standard deviation, and shape of the distribution).

Score		
94	66	76
32	78	32
7	11	86
36	74	56
40	73	87

3. Using the information from the previous two questions, select the appropriate analysis to determine if your class significantly differs from the class in question 1. Use a Type 1 error level of .05

4. After obtaining the previous results, you are now interested in determining how your classmates feel about other classes they are taking. Surprisingly, all of the students that you interviewed in your current class are also taking a physics course. Using the

398

information the students provide to you about their satisfaction with the statistics

course (data from question 2) and the information these same students provide about

their satisfaction with a physics course (below), determine if there is a significant

difference in the enjoyment ratings for the two courses.

Score		
43	47	27
54	23	41
26	25	35
48	28	15
29	16	29

5. A group of sports psychologists is interested in determining if people who run in the

Boston Marathon are younger than marathon runners nationwide. They believe that

this may be so because the Boston Marathon is considered to be one of the more

difficult marathons. The psychologists discover that the mean age of a marathon

runner nationwide is $\mu = 34$, with $\sigma = 7$. The average age of a sample of 36 runners

from the Boston Marathon was $M = 29$. Are those who run in the Boston Marathon

significantly younger than marathon runners nationwide?

6. You are interested in following up the research that was conducted in the previous

question by seeing if a group of your friends who are participating in a local marathon

are significantly different in age from those participating in the race overall. The

average age for the runners in your local marathon is $\mu = 31.4$, but you are unable to

obtain a measure of variability. The average age of your $n = 16$ friends is $M = 29.54$,

with $s = 7.98$. Does it appear that your friends are significantly different in age from

the runners in this race?

7. You are interested in determining if males and females differ in their preference of

cognitive therapy as a treatment for generalized anxiety disorder. You describe cognitive therapy to 15 males and 15 females with generalized anxiety disorder to gauge their level of preference for the treatment and obtain the scores provided below. Does it appear that there is a difference between males and females, using a Type 1 error rate of .05?

Females			Males		
12	12	11	8	8	6
12	11	6	6	7	10
9	6	8	10	6	5
8	12	11	5	4	6
11	14	13	6	5	10

8. After obtaining the result from the previous study (question 7), you are interested in whether ratings of the treatment will change after people are enrolled in the therapy. You decide to focus on the males and obtain the following data after they have completed treatment. Using the *male* data from question 7 as the pre-treatment condition and the data provided below as the post-treatment, does it appear that males' preference for cognitive therapy changes after treatment? (Participants appear in the same order location on both score lists.)

Pre-Treatment			Post-Treatment		
8	8	6	8	12	8
6	7	10	14	8	15
10	6	5	14	12	15
5	4	6	15	13	15
6	5	10	8	14	10

9. You are interested in comparing the influence of listening to various types of music while studying for an exam. You test 3 groups consisting of 10 individuals in each group. The first listens to rock music, the second listens to hip-hop, and the third is a control group that does not listen to music. The following are the test scores of all the individuals on the same English final exam. How does listening to music appear to influence test grades? Use any additional tests you need to discern the differences.

Rock	Hip-Hop	None
78	72	85
50	68	83
78	68	88
83	81	95
68	80	84
51	71	80
67	66	80
83	75	77
73	76	75
81	74	88

10. You are conducting research in the use of virtual reality in psychotherapy. You are interested in determining if those who have played video games respond differently to VR therapy than those who have not. You test a total of 50 participants, 25 who have played video games and 25 who have not. For the group that has played video games, you obtain the following data: $M = 36$, $\sigma = 17$. For the group that has not played video games, you obtain the following data: $M = 28$, $\sigma = 15$. Use a 5% Type 1 error rate to determine if there are any differences.

11. A team of researchers is interested in determining if a new osteoporosis medication is effective in increasing the density of a person's bones. The researchers compare the effects of the new medication to those of an established medication and to responses in a control group. The data are provided below. Determine if there is a significant difference in the effects of the medications on osteoporosis.

New Medication		Current Medication		Control	
63	45	50	61	30	41
51	47	60	59	33	47
59	64	65	63	33	17
67	47	67	56	45	17

12. The researchers from the previous question also suspect that the age of the participants may influence the effects of the medication. The researchers divide their sample into distinct groups, those under the age of 35 and those older than 35, as 35 is considered a cutoff point in the development of osteoporosis. Using the following data, determine if the new medication is as effective as a previous medication and if there appears to be an effect of age on response to the medication.

	New Medication	Current Medication
Below 35	63	50
Below 35	51	60
Below 35	59	65
Below 35	67	67
Above 35	45	61
Above 35	47	59
Above 35	64	63
Above 35	47	56

13. A study is comparing the effectiveness of a new medication (Drug A) for treating the symptoms of schizophrenia to the effectiveness of another medication (Drug B) and to a behavioral therapy. Using the following data on the number of symptoms shown by schizophrenics who are taking one of the three treatments, determine how the new medication compares to the previous medication and to behavior therapy at reducing the symptoms of schizophrenia. Use a Type 1 error rate of .01.

Drug A	Drug B	Behavior Therapy
22	18	37
23	36	30
28	19	29
16	37	31
22	36	33

14. An ice cream company is interested in determining if air temperature is related to the amount of ice cream it sells. Using the following data, determine if there is a relationship between ice cream sales (in thousands of gallons) and temperature.

402

Temperature	Ice Cream Sales
84	8
98	7
96	12
91	12
88	9
97	10
67	8
87	12
92	12
103	13

15. A teacher commonly brings candy for her class. She brings in dark chocolate and milk chocolate, and she expects that equal numbers of the 50 students in the class will take each type. She discovers that 35 students took milk chocolate and the remaining 15 took dark chocolate. Determine if the observed values are significantly different from what was expected, using a 5% Type 1 error rate.

Computational Answers

1. *Mean* = 54.47; s = 29.02 (using N) or 30.04 (using $N - 1$), and the distribution appears to be bi-modal.

2. *Mean* = 56.53, s = 27.24 (using N) or 28.20 (using $N - 1$), and the distribution appears to be either normal or negatively skewed.

3. The appropriate test is an independent-samples t test. For this test, $t = -0.19$, $p > .05$. Retain the null hypothesis that there is no difference in enjoyment between the classes.

4. The appropriate test is a repeated-measures t test. For this test, $t = -2.95$, $p < .01$. Reject the null hypothesis; students appear to enjoy statistics more than physics.

5. The appropriate test is a normal deviate test, for which $Z = 4.27$, $p < .05$. Reject the null hypothesis; those who run in the Boston Marathon are significantly younger than those who run in other marathons nationwide.

6. The appropriate test is a one-sample t test, for which $t(15) = 0.93$, $p > .05$. Retain the null hypothesis; your friends are not significantly different in age from those in the marathon.

7. The appropriate test is an independent-samples t test, for which $t(28) = 4.44$, $p < .01$. Reject the null hypothesis; females rate cognitive therapy more positively than do males.

8. The appropriate test is a repeated-samples t test, for which $t(14) = -5.50$, $p < .05$. Reject the null hypothesis; there is a significant difference in the males' preference for cognitive therapy after completing treatment, with their preference significantly increased.

9. The appropriate test is a one-way ANOVA, for which $F(2, 27) = 6.23$, $p < .01$. Reject the null hypothesis that there is no difference in scores depending on the type of music listened to. Tukey HSD $= 9.3$. The results indicate that those who did not listen to music did better on the exam than those who listened to music.

10. The appropriate test is an independent-samples t test, for which $t(48) = 1.76$, $p > .05$. Retain the null hypothesis; there is no significant difference in the treatment responses of those who have played video games from the responses of those who have not played video games.

11. The appropriate test is a one-way ANOVA, for which $F(2, 21) = 21.27$, $p < .01$. Reject the null hypothesis. Tukey HSD at a .05 error level $= 11.3$. The new medication is significantly more effective than no medication but does not have a significantly different effect from the current medication.

12. The appropriate test is a factorial ANOVA. There was no main effect for medication, $F(1, 12) = 1.88$, $p > .05$, and no main effect for age, $F(1, 12) = 2.08$, $p > .05$. There was no significant interaction between age and medication, $F(1, 12) = 1.50$, $p < .05$.

13. The appropriate test is a one-way ANOVA. Retain the null hypothesis; no treatment was more effective than any of the others at reducing the symptoms of schizophrenia, $F(2, 12) = 3.09$, $p > .05$.

14. The appropriate test is a correlation. There is no significant relationship between the amount of ice cream sold and temperature, $r = .44$, $p > .05$.

15. The appropriate test is a chi-square goodness-of-fit test, for which $\chi^2(1, N = 50) = 8$. Reject the null hypothesis; the numbers of students who took each type of chocolate were significantly different from what was expected.

True/False Questions

1. All parametric tests seek to establish causal relationship between two variables.

2. Descriptive methods seek to provide information about a large data set in the form of only a few numbers.

3. In a normal distribution, all measures of central tendency are equal.

4. The mode is the best way to describe the central tendency of a bi-modal distribution.

5. A Pearson r is used to describe the relationship between a ratio variable and a nominal variable.

6. The variance provides a measure of dispersion in linear units.

7. A z score tells the location of a score in a distribution in standard deviation units.

8. Inferential statistics enable you to better describe a set of data.

9. A one-sample t test is used when the population standard deviation is unknown.

10. An independent-samples t test is used when comparing two groups and all the population parameters are unknown.

11. A factorial ANOVA is used when you have two independent variables.

12. A chi-square test is used when comparing frequencies of nominal variables.

13. A Tukey post hoc test is considered more conservative than a Scheffé post hoc test.

14. Power is an important facet of hypothesis testing because it will tell you how large an effect you can expect.

15. A study compares how well shoe polish works on different types of leather shoes—sneakers, dress shoes, and sandals. If the researchers were to compare the effectiveness of the shoe polish using 15 different shoes within each category, they should analyze this data with independent-samples t tests.

True/False Answers

1. False	6. False	11. True
2. True	7. True	12. True
3. True	8. False	13. False
4. True	9. True	14. False
5. False	10. True	15. False

Short-Answer Questions

1. What is the difference between an inferential statistic and a descriptive statistic?

2. What is a standardized score? Give an example of a standardized score.

3. What is the purpose of a post hoc test?

4. How does an ANOVA (F ratio test) differ from doing multiple t tests? What is the advantage of using an ANOVA over multiple t tests?

5. Dr. Cho is interested in studying the impact of having grandchildren on the happiness of older people. He decides to compare happiness ratings of people over the age of 65 with 0, 1, 2, and 3 grandchildren. What type of analysis should Dr. Cho use for these data?

6. You are studying coping strategies in different family configurations. You are interested in determining if responses on a measure of coping differ between single-parent and dual-parent families. How would you go about analyzing these data?

7. After completing the study outlined in question 6, you hypothesize that the gender of the child in the family may affect the responses of coping. You decide to incorporate a second independent variable, gender of the child, into your analysis. How would your analysis change from the response in question 6?

8. You are interested in examining aspects of relationships. Recently, you have collected data on the number of favors one person in the relationship asks for and the amount of frustration experienced by the other person in the relationship. If you were interested in determining how these variables were related, what type of analysis would you conduct?

9. Continuing your relationship research, you now want to determine if there is a significant difference in the amount of time spent watching TV between the people in the relationship. You obtain a measure of how many hours each member of the relationship watches TV and want to determine if there is a significant difference. Keep in mind that in doing this research; you want to compare the people within each couple rather than comparing them to members of other couples. What type of analysis should you conduct?

10. A professor expects certain numbers of his students to obtain grades of A, B, and C. After the end of the term, the professor wants to determine if his estimation was appropriate. What type of analysis should he conduct to determine if he had the correct estimate?

11. A marketing company wants to determine which sport to include in its next advertisement. Its researchers ask a focus group to rate their interest in the product when they see a baseball, basketball, or soccer star in an ad. If the goal of this research is to determine which sport receives the highest ratings, what type of analysis should be conducted?

12. A researcher is interested in determining if obsessive thoughts or compulsive behaviors have a bigger impact on the functioning of those diagnosed with obsessive-compulsive disorder. To investigate this matter, the researcher finds a sample of individuals diagnosed with OCD and asks them to rate the impairment in their lives that they attribute to obsessions and to compulsions. The researcher hopes to compare the responses. What type of analysis should the researcher use?

Answers

1. Inferential statistics are used to learn about larger populations from samples. Descriptive statistics are used to describe a set of data.

2. A standardized score is a raw score that is transformed to a scale of a common metric. Two examples of standardized scores are z scores and t scores.

3. A post hoc test is used to determine which groups are significantly different from one another after obtaining a significant F ratio in an ANOVA.

4. An ANOVA maintains a consistent level of Type 1 error when comparing more than two groups, whereas multiple *t* tests would inflate the chance of committing a Type 1 error.

5. An ANOVA would be most appropriate.

6. An independent-samples *t* test would be most appropriate.

7. A factorial ANOVA now would be most appropriate.

8. A correlation analysis would be most appropriate.

9. A related-measures *t* test would be most appropriate.

10. A chi-square goodness-of-fit test would be most appropriate.

11. A one-way ANOVA would be most appropriate.

12. A related-samples *t* test would be most appropriate.

Multiple-Choice Questions

1. Which of the following is true of the normal distribution?

   a. Mean > Median > Mode

   b. Mode > Median > Mean

   c. Mean = Median = Mode

   d. Median < Mode < Mean

2. In a normal distribution, as a score moves _____ from the mean, the probability of obtaining such a score by chance _____.

   a. Further; is increased

   b. Further; remains the same

   c. Closer; is increased

   d. Closer; remains the same

3. Which of the following is most appropriate to describe frequency data on two variables?

   a. Means

   b. Variances

   c. Standard deviations

   d. Frequency tables

4. After doing poorly on a difficult test, Jack is afraid to tell his parents his grade. What information should he provide to his parents with his grade to put his grade in an appropriate context?

    a. A frequency table of the grades of all the other students in the class

    b. His grade in another course, in which he is excelling

    c. The mean and standard deviation of the class's scores on the test

    d. How much he loves them

5. You have a friend who consistently procrastinates and is doing poorly in school. Being a studious person yourself, you want to show your friend how increasing the amount of time studying is related to better grades. Which of the following analyses would be the most appropriate to prove your point?

    a. Correlation

    b. Chi-square test

    c. ANOVA

    d. Normal deviate test

6. You are asked to obtain food orders for a wedding reception that will be attended by 100 people. There are three food options for each guest. How should you best organize this information?

    a. Chi-square goodness-of-fit test

    b. Frequency tables

    c. Normal deviate test

    d. Histogram

7. A study on writing disabilities is attempting to discern if those with a writing disability will struggle with standardized tests in writing if those tests do not include a writing component. Which of the following would be appropriate to examine this hypothesis?

a. A normal deviate test comparing the scores on the test obtained by a sample of those with a writing disability to the population of scores with those with a writing disability

b. A repeated measures test comparing the scores on the test of those with a writing disability to their scores on a different test of a different subject matter

c. A *t* test comparing the performance on the test of those with a writing disability to those without a writing disability

d. A chi-square goodness-of-fit test that would determine if a certain number of those with a writing disability pass the test as compared to the amount expected to pass

8. A clothing company wants to determine if the sports team that is featured on a basketball jersey influences the sales of that jersey. The past 1000 jerseys sold were from 10 different teams. How should the company go about determining if the team named on the jersey influences the number of jerseys sold?

a. It should do a one-way ANOVA comparing the frequencies of jerseys sold for each team

b. It should do a factorial ANOVA comparing the frequencies of jerseys sold for each team

c. It should do a chi-square goodness-of-fit test comparing the frequencies of jerseys sold to the expected numbers of jerseys sold

d. It should do a chi-square test of independence, comparing the frequencies of jerseys sold to the expected numbers of jerseys sold

9. A teacher wants to determine if grades for her final exam have changed from last term's class to this term's class. How should she go about determining this?

   a. An independent-samples t test   c. A normal deviate test

   b. A related-samples t test   d. A correlation

10. A telephone company that wants to determine when people make the most calls obtains frequencies of calls made prior to 9 P.M. and calls made after 9 P.M. It also finds out the locations from which the calls were made, either the East Coast or the West Coast. If the company expects that there will be equal numbers of calls regardless of location or time, what test(s) should it use to analyze the data?

    a. A one-way ANOVA

    b. A factorial ANOVA

    c. A correlation

    d. A chi-square test of independence

11. A therapist is interested in determining if the number of therapy sessions people have is related to the amount of change in their depressive symptoms (as measured on a ratio scale). What type of analysis should the researcher use to address this question?

    a. A one-way ANOVA

    b. A factorial ANOVA

    c. A correlation

    d. A chi-square test of independence

12. When a researcher rejects the null hypothesis, he may have made a _____ error. (Note the following question does not have choices C and D)

    a. Type 1

b. Type 2

13. A researcher is investigating a neurotransmitter that is involved in the development of new white blood cells. To determine if this neurotransmitter does increase white blood cell development, she compares the concentration of white blood cells in rats without the neurotransmitter to the concentration of white blood cells in rats with the neurotransmitter. Which of the following tests is most appropriate for this design?

    a. Two-tailed, repeated-measures t test

    b. One-tailed, repeated-measures t test

    c. Two-tailed, independent-samples t test

    d. One-tailed, independent-measures t test

14. A school superintendent is determining which of the 5 schools in her district are performing poorly, as determined by a national reading test. He obtains all the students' scores from each school. How should he go about determining which schools are doing significantly worse than the others?

    a. A one-way ANOVA and post hoc test

    b. A factorial ANOVA

    c. A correlation

    d. A chi-square test of independence

15. Which measure of central tendency is most appropriate for positively skewed data?

    a. The mean

    b. The median

    c. The mode

    d. They are all equivalent

16. A Pearson *r* is considered to be

    a. A descriptive statistic only

    b. An inferential statistic only

    c. Both an inferential and a descriptive statistic

    d. Neither an inferential or a descriptive statistic

17. Which method is optimal to visually represent a continuous variable?

    a. A histogram

    b. A frequency table

    c. A table of means

    d. A scatterplot

18. In comparing the means of samples that contain different people who are matched on some criterion, you should use a(n) _____ to determine group differences.

    a. Related-samples *t* test

    b. Independent-samples *t* test

    c. Factorial ANOVA

    d. Correlation coefficient

19. What criterion is used to determine whether to use a normal deviate test or a one-sample *t* test?

    a. The knowledge of the population mean

    b. The knowledge of the population standard deviation

    c. The knowledge of the sample standard deviation

    d. The knowledge of the sample mean

20. You are interested in evaluating the validity of the statement "Absence makes the heart grow fonder." You decide to determine if the length of time people are apart is

related to the amount of affection they feel toward their partners. You find 40 couples in which the members have been apart from each other for varying amounts of time (continuously scaled) and decide to see if that amount of time is related the amount of affection (continuously scaled) the people have toward their partners. How would you go about analyzing these data?

    a. Related-samples *t* test

    b. Independent-samples *t* test

    c. Factorial ANOVA

    d. A correlation coefficient

Multiple-Choice Answers

1. C	10. D
2. C	11. C
3. D	12. A
4. C, if his parents understand statistics.	13. D
A if his parents do not understand	14. A
statistics.	15. B
5. A	16. C
6. B	17. A
7. C	18. A
8. C	19. B
9. A	20. D